T145 BE 2

c 4
FEB   - PAID

BASIC ELECTRONIC SWITCHING FOR TELEPHONE SYSTEMS

SECOND EDITION

DAVID TALLEY

HAYDEN BOOK COMPANY, INC.
Rochelle Park, New Jersey

JAN 1 2 1984

Library of Congress Cataloging in Publication Data

Talley, David.
    Basic electronic switching for telephone systems.

    Includes index.
    1. Telephone switching systems, Electronic.   I. Title.
TK6397.T33   1982   621.387   82-6182
ISBN 0-8104-0960-7   AACR2

Printed in the United States of America

2	3	4	5	6	7	8	9	PRINTING
83	84	85	86	87	88	89	90	YEAR

Preface

This revised edition of *Basic Electronic Switching for Telephone Systems* describes major advancements in electronic switching technology and equipment since the initial publication of this book. A new chapter on Digital Telephony addresses the transition from analog-to-digital switching techniques and the associated direct interconnection of digital transmission and switching facilities.

The fundamentals of electromechanical switching systems from the original direct control step-by-step method to the common control panel, rotary, and crossbar types are first succinctly reviewed. Then, the evolvement of solid-state devices and electronic modes, which led to the development of current electronic switching telephone systems, are further explained in simplified terms. Likewise, the essential principles of stored program control, central processing units, and the distributing processing means utilized in modern electronic switching systems are covered. The main objective is to provide a perceptive analysis of the major elements involved in electronic switching operations as well as the related transmission, signaling, and supervisory control functions.

In this respect, the basic design and functions of analog electronic switching systems, such as the Bell System's No. 1, No. 1A, No. 2, No. 2B, and No. 3 ESS, and the General System's No. 1 and No. 2 EAX are simply presented in the early chapters. The last chapter comprises descriptions of digital switching systems, particularly the Bell System's No. 4 ESS and the General System's No. 3 EAX, which are designed for toll switching. Descriptions are also included of the General System's GTD-5 EAX, the Nippon Electric NEAX-61, and the ITT System 12 types, which are intended for local or end-office applications.

This book has been written to furnish executives, engineers, technicians, and advanced students with a broad and in-depth knowledge of electronic switching precepts and the operations of contemporary electronic telephone central offices. The reader is referred to a previous book by the author, *Basic Telephone Switching Systems,* Second Edition, also published by the Hayden Book Company, Inc., for more detailed information on the telephony art, electromechancial switching systems, and interoffice signaling methods.

David Talley
Bal Harbour, Florida

Contents

(Electronic Automatic Exchange) • Call Control Operations in No. 1 EAX • Organization of No. 1 EAX Data Processor Unit • No. 2 EAX (Electronic Automatic Exchange) • Questions

Analog and Digital Signals' Fundamentals • Digital Switching Concepts • No. 4 ESS Toll and Tandem Digital Switching System • No. 3 EAX Digital Switching Toll Office • Requisites for Digital Switching End Central Offices • Digital Telephone Set Contingencies • Pair Gain System for Digital End Offices • Digital Switching Systems for Class 5 Offices • Questions

1

Telecommunication Switching Systems and Networks

Importance of Switching to Telecommunication Systems

A most important requisite of any telecommunication system, whether used for voice, data, teleprinter, television, or other forms of electrical communications, is its interconnection capability. The system must be able to connect quickly and accurately any station or terminal with any other one in the entire communications network.

These interconnections and switching networks have made possible the worldwide growth of telephone communications. A telephone system is normally comprised of one or more central offices to serve telephone customers in a town, city, or other locality. Several central offices serving the same local calling area are interconnected by a network of interoffice trunks. To reach telephones in other localities, whether in the same or distant states or in other countries of the world, the local central offices are connected through *toll centers* to national and international switching networks.

In the United States, a nationwide intertoll network designated Direct Distance Dialing (DDD) provides the facilities for handling telephone calls nationally and to most countries in the world. For instance, dialing seven digits will connect a telephone to any other one in its local area. Dialing ten digits will permit connections to any one of more than 185 million phones in the United States, Canada, and other countries on the North American continent. Moreover, international dialing has now been extended to over 82 countries including most of Europe, Central and South America, the Middle East, Japan, Australia, and New Zealand. Eventually, by dialing eleven to thirteen digits, one will be able to reach almost any telephone in the world. Table 1-1 lists the number of telephones in major areas of the world. It is anticipated that most of them soon will be reached by direct dialing.

Table 1-1.　Telephones in the World

Country	No. of Telephones in 1981	EMS*	ESS†
United States	175,505,000	64.7%	35.3%
Japan	55,421,515	99.7%	- - -
United Kingdom	26,651,384	74.4%	25.5%
West Germany	26,632,302	100.0%	- - -
Soviet Union	22,464,000	98.5%	- - -
France	22,211,952	93.6%	6.4%
Italy	18,084,996	100.0%	- - -
Canada	15,560,264	79.6%	20.1%
World	472,136,789	99.1%	

From "The World's Telephones," AT & T Co., 1981.
*EMS, electromechanical systems.
†ESS, electronic switching systems.

Functions of Central Offices

In a telephone system, the local switching center (usually designated the *end central office*) directly serves the telephones, dataphone sets, Picturephone® units, and other customer equipment connected to it. Calls destined for outside its service area are usually routed by the end central office to its associated toll switching center. This toll center will have access to the intertoll or DDD network of the Bell System and to international and global networks as later explained. Figure 1-1 diagrams the switching arrangements between central offices and a toll center.

The local or end central office contains various equipment for handling originating calls from customers and, likewise, for completing incoming calls to its telephone stations. It directly receives the digits dialed by the call-originating station, translates them, performs the required switching actions, and transmits details of the called number to the distant central office or toll center as may be required. End central offices also furnish transmission and signaling facilities including ringing, charging for calls, maintenance, and testing features. The application of electronic switching to such central offices and toll centers is the subject of this book.

The intertoll network comprises toll centers and primary, sectional, and regional centers. These switching centers, as subsequently explained, interconnect trunks for routing calls to their destinations but have no direct access to telephone stations. In handling a toll or long-distance call, the originating central office transmits the ten-digit number of the called party (three-digit area code plus seven-digit directory number) to its associated

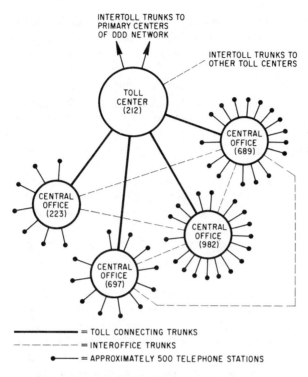

Fig. 1-1. Local Telephone Switching Networks

toll center. The first three digits (area code) or the first six digits (area code plus central office code) of the called number are translated by the toll center, depending on the call's destination, routes available, and other factors. The toll switching center then performs the necessary switching functions to route the call toward its destination.

Telephone Numbering Plans

To provide for local, national, and international or global communications, it is essential that every telephone station be uniquely identified, as are social security numbers. Consequently, a seven-digit telephone number has been standardized in the United States and many other countries. The first three digits make up the central office code and the last four contain the line number. For example, 555 is the central office designation or office code of telephone number 555-1212. On the basis of a four-digit line number, each central office will have a theoretical numbering capacity of 10,000 lines in the sequence 0000 to 9999. (It may be of interest that about 33,000 central offices and over 100 tandem centers were required to serve the approximate 176 million telephones in the United States in 1981. This included, in addition to the Bell System, the telephones served by the General System, United Telephone System, and various other independent telephone companies.)

It is believed that the telephone numbering plan started with the manual central office switchboard. The largest switchboard had a capacity of 10,000 line jacks which were numbered from 0000 to 9999. In cities and other localities where additional central offices were needed, each four-digit line number was preceded by the designation of its central office, usually named after the locality or a historic personality.

With the advent of dial automatic switching, this numbering scheme initially was continued in the larger cities, except that only the first three letters of the central office name were dialed before the line number. In the smaller towns and cities, one or two digits were originally used as the office code because the electromechanical switching equipment was of the step-by-step type and required a separate switching level for each digit dialed.

The rapid growth of telephone service, particularly in urban areas, began to deplete the number of office codes derived from the first three letters of the central office name. Moreover, only 330 codes were found usable from the possible 1000 (10 x 10 x 10) combinations of the three digits. As a result, the assignment of two letters, and a numeral in place of the third letter, was next introduced. This plan increased to 576 the total number of usable central office codes. The continued demand for telephones, however, led to the present all-number calling plan (ANC) whereby numbers replaced all letter combinations for office codes. A total of 800 (8 x 10 x 10) possible combinations is provided by ANC because the numerals 1 and 0 are not usable for the first digit of an office code. Because of the requirements for geographical area codes and special-purpose codes, such as 411 for information and 911, the national emergency code for police assistance, only 598 central office codes are available for assignment. Considering a maximum of 10,000 numbers per central office code, it is possible to have 5,980,000 telephone numbers in each area code region.

For long-distance dialing purposes utilizing the DDD network, the Bell System, with the cooperation of other telephone companies, has divided the United States and Canada into approximately 125 telephone regions, each assigned a three-figure area code. The specific area code is dialed only if the called telephone number has an area different from that of the calling telephone number. (In areas with step-by-step (S × S) central offices, it is necessary to dial the digit 1 prior to the area code.)

The three-digit area code is added to the assigned seven-digit directory number to make a total of ten digits to be dialed. The first digit of the area code does *not* contain the numbers 0 or 1. The second or middle digit *must* be either 0 or 1, and the last digit utilizes *all* numerals from 1 to 9 and 0. This area code numbering plan provides a total of 160 (8 x 2 x 10) possible codes. The 0 and 1 designations for the middle digit enable the central office switching equipment to distinguish a toll or long-distance call from a local area call. Some typical area codes are listed in Table 1-2. Area codes are used to route calls from one telephone switching network area to another but are not used within the same network area.

Table 1-2. Area Code Numbering Plan

Location	Area Code
Boston	617
New York	212
Miami	305
Chicago	312
Washington, D.C.	202
Dallas	214
Los Angeles	213
San Francisco	415
San Juan, P.R.	809
Montreal, Canada	514
Mexico City	903

Direct Distance Dialing Network

The Bell System's Direct Distance Dialing (DDD) network has made possible the interconnection of telephone stations in the United States, Canada, Mexico, and almost every country on the North American continent. This network is controlled by switching centers designated, in order of importance, *regional* (Class 1), *sectional* (Class 2), and *primary* (Class 3) centers. These three centers are connected to the *toll centers* (Class 4 offices) referred to in Fig. 1-1. A simplified representation of the DDD network is illustrated in Fig. 1-2.

End central offices (Class 5 offices) connect to the DDD network through the toll centers. Calls between local or end central offices (Class 5) in the same area code region may be routed over interoffice trunks or through a tandem office as shown in Fig. 1-2. Calls through toll centers normally are sent over direct high-usage routes. If these routes are busy or not available, alternate switching routes through primary, sectional, and regional switching centers are selected in that sequence.

The aforementioned switching centers in the DDD network are interconnected by coaxial cables, point-to-point microwave radio systems, and satellite circuits. To provide the many thousands of voice and data channels, carrier (multiplex) transmission of signals is used on these coaxial cable and the microwave and satellite radio facilities.

Referring to Fig. 1-2, let us follow the routing of a telephone call from Boston to Los Angeles. Dialing area code 213, followed by the seven digits of the called party in Los Angeles, will cause the originating central office

Fig. 1-2. Simplified Direct Distance Dialing Network

(Class 5) to switch the call to its associated toll center (Class 4). The switching equipment in this toll center, upon receiving the 213 area code, will immediately pass the call to the primary center serving the Boston area. The primary center (Class 3) first will try to complete the call over the high-usage trunks designated 1, 2, and 3 in Fig. 1-2. If routes 1 and 2 are busy, the call will be sent over route 3 to the sectional center (Class 2) in Los Angeles. This Class 2 office, consequently, will switch the call via final route trunks through the primary (Class 3) and toll center (Class 4) offices serving the Los Angeles area to reach the particular end central office (Class 5) serving the called party.

If the outgoing intertoll trunks from the Boston primary center (Class 3) to Los Angeles are busy, route 4 to the sectional center will be selected. The switching equipment in this Class 2 office will attempt to complete the call over routes marked A, B, or C before selecting final route D to the regional center (Class 1). Regional centers are interconnected by direct final route trunks. Therefore, the call can be routed from the Los Angeles regional center (Class 1) to the desired end central office (Class 5) through the respective sectional, primary, and toll centers shown in Fig. 1-2.

Worldwide Dialing

Worldwide or International Direct Distance Dialing (IDDD) of telephone calls and other communication interconnections has been substantially advanced by recent developments in the communications field. These improvements include the installation of high-capacity and repeater-type telephone cables under the oceans, and wide-band communication satellites in synchronous orbits around the earth. These new facilities provide many thousands of channels for voice, data, and related communication needs.

To facilitate the use of these global communication systems, a world zone-numbering plan has been formulated by the International Telegraph and Telephone Consultative Committee (CCITT) of the International Telecommunications Union. This plan divides the world into nine telephone numbering zones as illustrated in Table 1-3.

To make an international call over the worldwide IDDD network, a customer in the United States first must dial the international access code "011" for station-to-station calls, or "01" for person, collect, or credit card calls that require the assistance of an operator. The customer next dials the assigned two- or three-digit *country code* followed by the *city routing code* and then the desired local number. The countries in each world telephone zone are assigned a country code comprising two or three digits, the first digit being the respective world zone code. Nations with many telephones and national numbers of 10 digits or less are assigned two-digit country codes. Those with fewer telephones and national numbers of nine or less digits are given three-digit codes. This procedure will limit the number of digits to be dialed following the international access code.

Some representative country codes are: England, 44; France, 33; Italy,

Table 1-3. World Telephone Numbering Plan

World Telephone Zone Code	Zone or Region
1	North America including Hawaii and Caribbean area except Cuba
2	Africa
3	Europe
4	Europe
5	South and Central America including Mexico and Cuba
6	South Pacific including Australia
7	U.S.S.R.
8	Far East including Japan
9	Middle East and Southeast Asia

39; Sweden, 46; Israel, 972; Hong Kong, 852; Japan, 81; Australia, 61; Peru, 51; and South Africa, 27. The city code consists of one to three digits representing the major cities in each country. Its function may be compared to that of the area code in the United States. For example, a person in Chicago would dial the following sequence of digits to reach a particular telephone in Paris: international access code "011" + country code for France "33" + city code for Paris "1" + local number "555-1212."

Consequently, the complete international dialing sequence, or *address information*, may be as long as 14 digits. Local (Class 5) central offices and the DDD network, however, normally can accept a maximum of 11 digits. This problem has been solved by providing two-stage signaling of address information in central offices handling international calls. The international access prefix, "011" or "01," when received by the central office switching equipment, causes a connection to be made to an electronic switching No. 4 ESS unit serving as the international toll or *gateway office*. Then, the complete international called number (maximum of 11 digits) is recorded by the central office and transmitted to the No. 4 ESS *gateway office* over the aforementioned connection for subsequent processing.

Telephone Station Equipment

The telephone instrument (commonly called the *subset*), used at customers' stations to initiate or answer calls, is a sophisticated and reliable device designed to function with different switching systems. It normally consists of a handset containing the receiver and transmitter units (some types may include a rotary or pushbutton dial), and a small apparatus box containing the ringer, transmission circuit, rotary dial, or pushbutton keyset, and the hookswitch for the handset. Typical telephone instruments are shown in Figs. 1-3 and 1-5. The telephone subset performs these general functions:

1. Signals an incoming call by the bell-ringing signal
2. Signals the central office equipment whenever a call is originated or answered (handset off-hook), and terminated upon disconnection (handset on-hook)
3. Transmits dial pulses or pushbutton tones (address signals) to the central office equipment
4. Enables two-way operation (transmission and reception) over the two-wire telephone line
5. Converts speech to electrical signals and vice versa through the handset.

Telephone subsets also may be designed for four-wire operation which requires separate pairs of wires for each direction of transmission. Because of economic considerations, a single pair of conductors is employed to connect telephone stations to the central office on a two-wire transmission basis. Four-wire circuits, however, are normally employed for toll and long-distance trunks and in the nationwide DDD network. Likewise, interoffice and short-haul trunks, which utilize carrier or multiplex facilities, use separate pairs for each transmission direction to provide four-wire operation.

Rotary Dial and Pushbutton Dialing

The previous references to dialing of telephone numbers include the use of either the rotary dial or pushbutton dialing by the telephone user. Figure 1-3 is a photograph of a typical telephone instrument or subset equipped with a rotary dial. The rotary dial, which initially was developed in the early days of automatic switching equipment, is essentially a circular device with ten finger-holes numbered 1 to 9 and 0, arranged in a counter-clockwise direction around its periphery. The rotary dial opens and closes contacts as it spins back to its normal stop position, from the digit to which it was turned, thereby generating *direct current* (dc) pulses. These pulses, which correspond to the numeral of the selected digit, are used as "address" signals for instructing central office switching equipment regarding the called telephone number.

Fig. 1-3. Telephone with Rotary Dial (Courtesy of AT&T Co.)

Fig. 1-4. Pushbutton Dialing Frequencies

Rotary dials generate pulses at the rate of 10 per second irrespective of the digit dialed. Dials on switchboards of private branch exchanges (PBX) or private automatic branch exchanges (PABX) can send 20 pulses per second, thus reducing the operating time of handling calls.

Some 12 or more seconds are normally required to dial the seven digits of a local telephone number, and at least 14 seconds to dial ten digits when making toll or long-distance calls. This time period has proven satisfactory for electromechanical switching offices, particularly the step-by-step type which function directly from dial pulses as sent from the telephone dial. The aforementioned dial-time periods mean that certain common equipment in the central office, such as the register-sender, will not function until all the dial pulses are received. It is apparent, therefore, that the longer the dialing period, the greater the amount of common control and related switching equipment that will be needed to handle the many simultaneous calls during busy-hour periods.

The very high switching speeds made possible by electronic switching systems have stressed the need of more rapid and accurate dialing means by telephone users. Key pulsing, a form of multifrequency (MF) pushbutton calling using combinations of two tones, had been used for many years by operators on the old manual toll switchboards and the present traffic service position system (TSPS) consoles. This particular multi-frequency dialing system is not suited for use with telephone instruments because of the need to provide filter circuits to guard against voice interference and other considerations. As a result, a new pushbutton dialing method designated *Touch Tone®* had been developed by the Bell Telephone Laboratories for use with electronic switching systems. It is also usable with present electromechanical central offices by providing suitable tone-receiving and conversion equipment. Touch Tone® dialing, also known as touch calling, utilizes eight frequencies in the 700–1700 Hz range divided into low-band and high-band tones, as illustrated in Fig. 1-4. These eight frequencies, selected to avoid harmonically related interference from voice signals, are designed for a maximum of sixteen pushbuttons containing four rows and four columns on the faceplate of the pushbutton dial. A typical pushbutton or touch calling telephone set in current use is shown in Fig. 1-5. The twelve pushbuttons are arranged in four rows and three columns as illustrated. Pushbuttons designated * and #, corresponding to digits 11 and 12, are primarily for use with electronic switching offices. The functional circuit of the GTE Automatic Electric touch calling keyset is represented by the schematic diagram of Fig. 1-6. Transistor Q1 and its associated components—inductors L-3 and L-4. capacitor C-3, resistors R-2 and R-3, and diode D-1—make up an *audio oscillator*. The audio frequencies produced by this circuit are controlled by the operation of the keyset pushbuttons in accordance with the frequency plan outlined in Fig. 1-4.

The following example explains the operations in a keyset unit that utilizes inductance and capacitive elements to generate the audio tones. Assume that digit 7 (PRS), located in the third row of the first column, is operated. Referring to Fig. 1-6, this action will close row 3 contacts of switch S-1 and capacitor C-1 will be connected to the third tap on inductor

Fig. 1-5. Pushbutton Telephone (Courtesy of GTE Automatic Electric Co.)

Fig. 1-6. Schematic Diagram of Touch Calling Keyset

L-1, thereby establishing a resonant circuit for producing the 852-Hz tone. At the same time, another resonant circuit for creating the 1209-Hz tone will be constituted by the closure of the S-2 switch contacts in column 1 to connect capacitor C-2 to the first tap on inductor L-2. Switch S-3, which is common to both the S-1 and S-2 switch contacts, cannot operate until the pushbutton has been fully depressed. When the contacts of common switch S-3 are closed, a low dc voltage (produced from the diode rectifier unit bridged across the telephone line) will be connected to activate transistor Q1 and cause the generation of the 852- and 1209-Hz audio tones. RV-1 and RV-2 are semiconductor rectifiers, termed varistors, which are employed to reduce interactions between the respective adjacent frequencies. The receiver (REC.) in the handset will be shunted by the 5100-ohm resistor R-1 whenever switch S-3 is operated, but the calling party will be able to hear the transmitted tone signals corresponding to the dialed or keyed digits, although at a reduced level.

An important advantage of pushbutton or Touch Tone® dialing is that the audio tones generated for each digit are in the voice-frequency range. Therefore, they can be transmitted anywhere over the DDD switching network in the same way as speech signals. This outstanding characteristic can be utilized to provide many new services for telephone customers. For instance, a customer may dial the number of an automatic department store, and place an order by dialing the digits corresponding to his or her account number and to a coded list of desired items. These services are not possible with the rotary dial because the dc pulses normally produced control only the equipment in the local central office. Furthermore, almost all interoffice and toll trunks employ carrier transmission so that metallic paths are not available for handling dc dial pulses.

Additional equipment is required in electromechanical central offices to receive and translate tone signals from pushbutton dials. For step-by-step offices, the touch calling signals are converted into direct current dial pulses to permit direct operation of the selector switches. In common control offices, such as the No. 5 Crossbar, the Touch Tone® signals are detected and translated into the proper form required by the given switching equipment.

Automatic Switching Systems

Before discussing electronic switching systems, it is desirable to understand the concepts involved in the development of automatic telephone switching, and the operations of major electromechanical switching systems. Automatic switching equipment initially was designed to aid and later to replace a telephone operator and a manual switchboard. The evolvement of electromechanical systems has simulated to a large degree the switching functions of the cords, plugs, and jacks on manual switchboards. These functions performed by the operator in processing a local call on a manual switchboard are as follows:

1. The operator scans the switchboard panels for lights indicating an originating call.
2. When a lamp lights, the operator immediately notes its location, the identity of the calling line's jack, and records this information.
3. The operator inserts the plug of an answering cord into the jack of the calling line and announces, "Operator."
4. The called line number is received from the calling party and recorded by the operator.
5. The jack of the called party on the switchboard is located.

6. The operator determines the state of the called line by testing the sleeve of its jack with the tip of her calling cord's plug.
7. If the called line tests idle (click is not heard by the operator), the calling cord's plug is inserted into the called line's jack to complete the connection. Ringing is started automatically by the relay equipment in the cord circuit.
8. If the called line tests busy (click is heard by the operator) the plug of the calling cord is inserted into a busy-tone jack which transmits a busy tone to the calling party.
9. When the connection is completed, per steps 7 or 8 above, the operator is available to handle another call.
10. The progress of the call is indicated by lamps on the keyshelf of the switchboard which are associated with each answering and calling cord.
11. The operator releases the connection by removing the answering and calling cords when both lamps light to indicate that the calling and called parties have disconnected. The operator can perform this function while processing other calls.

Automatic or electromechanical switching equipments do not *exactly* duplicate the direct interconnections completed by the operator using the cords, jacks, keys, and other switchboard equipment. Relays, vertical and rotary motions of stepping switches, brush-type selectors, and crosspoint contacts of crossbar switches are utilized for interconnecting telephone lines in electromechanical central offices. Note that central offices, whether manual, electromechanical, or electronic switching types, are not designed for the simultaneous connections of all or even a majority of their customers. For instance, to permit all customers to talk to each other at the same time, the number of required interconnections in the central office would amount to one-half the number of customers. As a rule, only a small portion of customers, normally 10 percent to 14 percent, simultaneously originates calls within a given time of busy-hour periods. Consequently, the central office equipment needs the capacity to serve only that percent of the total lines. In a central office serving 6,000 lines, for example, approximately 700 calls would usually be connected or in progress of completion at any one instant in busy-hour periods. Thus, only the amount of equipment capable of handling that number of simultaneous connections need be provided. This principle is also followed in the design of electronic switching central offices.

Direct Control Step-by-Step Switching Systems

Electromechanical switching systems are usually divided into two general categories: *step-by-step* (S × S) *direct control* and *common control* systems. The original direct control switching method evolved from the invention of the Strowger switch used in the initial automatic dial system. It would advance step-by-step in unison with the dial pulses. The S × S method od may be defined as a direct progressive control system in which stepping-type switches are directly controlled by pulses from customer's dials. As a result, there is an established correlation between the speed of the dial pulses and the motion of the stepping switches in the central office. The S × S switching technique is still used today in many electromechanical central offices.

Figure 1-7 is a block schematic drawing which shows the principal elements of the switch train in an S × S central office and their relations with the dialed digits of a four-digit line number. The stepping or selection switches are designed for both vertical and horizontal movements which are called, respectively, the selection and hunting stages. These switches can step a maximum of ten levels vertically in unison with dial pulses and then, in the hunting stage, automatically rotate horizontally over ten terminals to search for an idle path to the next selector or other circuit. The time interval between the digits dialed by customers is the factor limiting to ten the number of horizontal terminals provided in the hunting stage. The last selecting switch is called the connector because it connects directly to the terminals of the called line number. The connector switch has two selection stages but does not hunt. The vertical stepping is controlled by the tens digit and the horizontal rotation is controlled by the units digit of the called line number.

Fig. 1-7. Direct Control S × S Switching System

To illustrate the processing of a local or intraoffice call within a small S × S central office, let us assume that only four-digit line numbers are assigned, and that the number 4356 is to be called from a customer's telephone. As we see in Fig. 1-7, when the calling party initiates the call (handset off-hook), an idle linefinder switch (always associated with a first selector) finds and connects to the calling line. A dial tone is sent to the calling party by this first selector. Dialing the thousands digit 4 steps the first selector vertically to its fourth level. Then the selector rotates horizontally to select an idle path to a second selector that serves the 4000 number series. The next digit, 3, will step the second selector to its third level where it will hunt for an idle path to connector switches associated with lines in the 4300–4399 number sequence. The next or tens digit 5 will cause the connector switch to move to its fifth level. Since this switch has no hunting feature, the last or units digit 6 will step this connector horizontally six positions, thereby connecting to the called line number 4356. The connector next tests the called line; if it is idle, ringing current is applied. If the line is busy, a busy tone is returned to the caller.

Direct control switching systems, such as the step-by-step, have a number of disadvantages, however, especially when several central offices are required in the same local service area. A prime drawback is that the office code digits define the exact levels to which selectors are stepped. For example, office code 637 will cause the first selector to step to the sixth level, the second selector to the third level, and the third selector to its seventh level. It is not possible to change this fixed relationship between central office codes and locations on selector switches of interconnecting paths or outgoing trunks to other central offices. Once the customer has dialed the three digits of an office code, it is not feasible to select an alternate route if all paths or trunks to the called central office should be busy. In such cases, a busy signal will be sent back to the calling party.

Another disadvantage is that the tens and units digits of the called line number specifically determine its location on the connector switch. Traffic overloads, therefore, could arise if numerous customers who receive many calls are assigned line numbers within the same hundreds group, for example: 3100–3199, 6700–6799, 8000–8099, etc.

Common Control Switching Systems

For maximum efficiency, switching systems theoretically should have full access and nonblocking capabilities. *Full access,* however, actually means that every telephone line can be connected to every other telephone but not necessarily at the same time. *Nonblocking* means that any originating call in a central office can be connected to any other telephone irrespective of the number of existing connections. These goals are substantially obtained in most common control switching systems.

Common control switching systems of the electromechanical type in current use include some panel and crossbar categories manufactured by Western Electric Company for the Bell System, and the rotary non-Bell or director systems which are used by independent telephone companies. The director system, which was developed by GTE Automatic Electric Company, applies some of the main features of common control operations to step-by-step central offices. This common control switching method is used by a number of General System and other non-Bell telephone companies.

Two main characteristics of common control switching systems are the employment of large capacity switches or switch frames, and use of groups of common control circuits commonly termed *register-senders, originating senders,* or *senders,* for temporary storage and translation of received dial pulses into appropriate signals for controlling the switching network. For instance, the switch frames in panel dial central offices have access to 500 levels. In the crossbar system, a combination of crossbar switches can provide access to several thousand outlets. On the other hand, selectors in S × S offices have a capacity of only ten levels, with ten paths on each level, or a possible 100 outlets.

Moreover, in common control offices, the switching operations are separated from the actions controlling them. This permits a particular group of common control circuits, such as register-senders, to direct the routing of connections through the switching network for many calls at the same time. Furthermore, as soon as a particular register-sender completes the routing of a call to its destination, it releases and becomes immediately available to handle other calls.

Panel Dial System

The panel dial system, now obsolete, initially was developed in the 1920s by the Bell System to replace manual central offices in the larger cities. It was a common control, progressive switching type. It had five major categories of vertical selector switching frames: *linefinder, district,* and *office* for handling outgoing calls; and *incoming* and *final* frames for completing connections to the called line number. Two principal common-control devices were also utilized, namely, the *register-sender* and the *decoder* or *decoder-marker* circuits. The register-sender provided a dial tone to the calling line, temporarily stored the called line number as dialed by the calling party, and connected to a decoder circuit. The decoder translated the first three digits (area or office code) into proper routing instructions which were sent back to the register-sender. The register-sender proceeded to control the selection operations of the switching network, consisting of the district, office, incoming, and final frames. A brief description of the equipment used in the panel system and its call-processing operations follows. (Portions of the balance of this chapter are adapted from the author's *Basic Telephone Switching Systems,* Second Edition, published by Hayden Book Company in 1979.)

Linefinder frames usually were equipped with ten vertical banks of forty line terminals per bank or 400 per frame. The other selector frames had five vertical banks, designated 0 to 4 inclusive, each having 100 sets of three terminals corresponding to the standard tip, ring, and sleeve conductor terminology. These terminals protruded through both sides of the banks for contacting the brushes on selectors. Each selector frame had a capacity of sixty vertical selectors driven by cork rollers, one for updrive and the other for downdrive or restore to normal. These cork rollers were continuously rotated by electric motors, two on each selector frame. An eighteen-position rotary switch, called a sequence switch, both was associated with and controlled the operation of each selector on the frame. This switch was rotated to different positions by a revolving vertical drive shaft, operated by the earlier-mentioned electric motor, under control of an electromagnetic clutch and related relay circuits. Various circuit combinations were made or changed as the sequence switch moved to different positions, thus controlling the diverse functions in handling a call.

Figure 1-8 is a simplified block diagram of a panel switching system. Let us follow a local interoffice call through the switching network. When the calling party lifts the handset, a linefinder is used to search for the calling line's terminals on a particular bank of the linefinder frame. At the same time, a district-sender link circuit functions to select an idle register-sender which will send a dial tone to the calling telephone. Assuming that the number 637-7890 is dialed, as soon as the first three digits (637) are received by the register-sender, it will connect to a decoder and transmit the digits 637. The decoder translates this office code into the particular brush and group selections to be made by the district and office selectors involved in the call, returns this information to the register-sender, and immediately releases. The register-sender now proceeds to direct the district selector to a specific bank and group of terminals to find an idle path to an office

Fig. 1-8. Panel Dial System

selector. It then directs the chosen office selector to the particular bank and group location of the outgoing trunks to central office 637. The office selector hunts for an idle trunk in this goup of terminals, thereby completing the routing instructions that had been received by the register-sender from the decoder.

The interoffice trunk selected by the office selector terminates on a selector circuit in an incoming frame of panel office 637. The register-sender, having previously recorded 7890 as the called line number, directs the incoming selector to trip its brush 3 and then advance to group 3, the top group of twenty-five terminals on bank 3 of the incoming frame. This particular group of bank terminals has access only to the final frame selectors which can connect to line numbers in the 7500–7999 series. The register-sender next commands the final selector to trip its brush 3 and then travel to terminal 90, thereby connecting to line number 7890, which is assumed to be idle. This completes the functions of the register-sender and it releases. Simultaneously, the district selector circuit cuts through the talking path from the linefinder to the interoffice trunk, and the incoming selector circuit applies a ringing current to the called line. If the called line had tested busy, the final selector would have returned to normal and a busy tone would have been returned to the calling party. When the calling party hangs up, the district, office, incoming, and final selectors disconnect and restore to normal in that order.

Rotary Switching System

The same design principles of the panel system are followed in the rotary switching system, except that different types of power-driven switch frames are employed. The rotary system, as its name implies, utilizes rotary selectors on semicircular banks for switching operations. It includes similar power-driven sequence switches and common control equipment, such as register-senders, decoders or translators, and linefinders, as used in the panel system.

Trial installations of both the rotary and panel systems were originally made in the early 1920s, but subsequently, the Bell System decided to standardize the panel system for the large cities in the United States. The rotary system has been substantially improved over the years, especially by the Bell Telephone Manufacturing Co. of Antwerp, Belgium, which is affiliated with the International Telephone & Telegraph Corp. Installations of the types 7-A to 7-E Rotary System have been made in many countries of Europe and other parts of the world. Some rotary systems have been installed for independent telephone companies in the United States.

Common Control Step-by-Step System

The rapid growth of telephone requirements in small communities, the advent of the Direct Distance Dialing network, and extended area service (EAS) have necessitated the assignment of standard seven-digit numbers to all customer telephones served by step-by-step (S × S) central offices. These factors and the need to provide alternate routings for outgoing traffic in S × S offices, accentuated the necessity to apply common control operations to existing S × S offices. The techniques for this purpose had been known for many years, but had not been previously considered economical or essential for small S × S central offices.

One method for incorporating major common control features into existing S × S installations without modifying the selector switches and switch train is the 101 Director System developed by GTE Automatic Electric Inc. It provides register-senders with associated access relay equipment, and an electronic translator which may be common to as many as 100 register-senders. This translator is a very high-speed electronic device which can receive and translate a maximum of six digits (area and office codes), and provide routing instructions to the register-sender for completing the call. A block diagram showing these common control techniques as applied to a typical S × S office is presented in Fig. 1-9. Note that the access circuit for selecting a register-sender is inserted between the linefinder and first selector in the switch train.

On local and intraoffice calls, the first two digits of the office code usually must be absorbed because only the third digit is required to direct the first selector to the proper level for selecting an idle path to the next selector in the switch train. For example, assume the number 637-4321 is dialed. The register-sender immediately repeats the first digit 6 to the first selector, and also to an idle electronic translator. The translator, recognizing that the first digit 6 may signify a local or intraoffice call, orders the register-sender to release the first selector. When the second digit 3 is dialed, the code 63 is sent to the translator which, knowing it as the elements of a local call, directs the register-sender to absorb these two digits, and to repeat the next dialed digit to the first selector. Therefore, as soon as the third digit 7 is dialed, the register-sender repeats it to the first selector, as was ordered by the translator. At the same time, the complete office code digits 637 are sent by the register-sender to the translator, as an indication of completion of office code dialing. The translator will now cause the release of the register-sender. The subsequent dialing of the called line number 4321 will directly control the operation of the other selectors in the switch train in the usual manner.

On interoffice calls when all office code digits are required for routing the call, the register-sender can provide alternate routings. For instance, as-

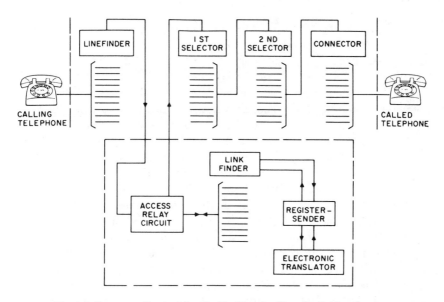

Fig. 1-9. Common Control Applied to Step-by-Step Switching System

sume that all ten trunks to a certain office, which appear on level 4, second selector switch, are busy. The register-sender, upon receiving this indication, and knowing that an alternate route is required in accordance with instructions previously received from the translator, will release the first and second selectors by momentarily opening the dialing loop (tip and ring leads). Then, the register-sender will direct the first and second selectors, in sequence, to other levels in accordance with instructions received from the translator, in order to select an idle trunk in the alternate route.

Crossbar Switching Systems

The panel dial and rotary switching common control systems have numerous disadvantages. Among them are the use of power-driven selectors and sequence switches, sliding contacts on bank terminals which are a source of noise, and the comparatively long time required by selectors to find idle paths and to complete calls. Moreover, considerable equipment, including sequence switches and relays directly associated with each selector, are used for only a few seconds during the establishment and disconnection periods of a call. As a result, substantial quantities of apparatus are needed in such common control central offices to handle normal traffic volume. These factors and the high cost of power-driven mechanisms led to the development by the Bell Telephone Laboratories of the crossbar switching system utilizing the crossbar switch.

The original crossbar switching design was the No. 1 Crossbar System which the Bell System installed during the 1930s in New York and other large cities. The switching network of the crossbar system is not bound in any manner by the called number assignments. Customer line numbers have only a single appearance on a line-link frame, which handles both originating and terminating calls, in contrast to connecting to both a linefinder and final frame in the panel system. The switching network employs crossbar switches, each normally having 200 sets of crosspoint contacts in twenty horizontal and ten vertical rows on the bank. Each crosspoint actually has twin precious-metal contacts which may be pressed together as in the operation of a relay, thus minimizing sliding contacts.

The principal control device of the crossbar switching system is the *marker circuit* which, in addition to translating dialed digits as the decoder does in the panel system, locates idle paths and trunks for routing calls through the switching network. If first-choice paths through the network are busy, the marker makes a second attempt. Similarly, the marker will select an alternate route if all direct trunks to the called central office are busy. The marker also determines whether the called line number is idle or busy. Consequently, the marker may be compared to a computer because it receives information, refers to its memory for programmed instructions, and then initiates or directs other equipment to take appropriate actions in processing and completing calls.

The simplified block diagram in Fig. 1-10 illustrates the basic elements of the No. 1 Crossbar System. Separate markers and switching networks are employed for originating and terminating traffic, respectively. The number group frame indicated in Fig. 1-10 tells the terminating marker on which particular line-link frame and at what specific crossbar switch crosspoint it will find the line terminal corresponding to the called directory number. The line number on the line-link frame need not necessarily correspond to the directory number. This permits greater flexibility in handling customer traffic loads because, for instance, it is not necessary that all private branch exchange (PBX) lines have consecutive directory numbers as in other electromechanical switching systems.

The NO. 5 Crossbar System, which was initially placed in service by the Bell System in 1947, is probably the ultimate in sophisticated electromechanical switching systems. Unlike the No. 1 type, it utilizes one switching network and a single marker to handle both originating and terminating traffic. Figure 1-11 is a block diagram showing the major components of the No. 5 Crossbar System, including the combined switching network (line-link and trunk-link frames) and the marker with its associated register and sender circuits. The connector circuits in this diagram are used to connect the marker with other common control equipments, and to the line-link and trunk-link frames. These connecting paths are utilized only for about 0.2

second during the progress of a call. A synopsis of the processes involved in handling an interoffice local call follows.

A customer lifting the handset to initiate a call with cause the operation of the line (L) relay on a particular line-link frame, which in turn activates the associated connector circuit to seize an idle marker. The marker proceeds to locate the calling line on the particular line-link frame, and secures an idle originating register on a trunk-link frame. It then closes a selected path from the calling line terminals on the line-link frame, through the junctor group and trunk-line frames, to the originating sender before releasing. The originating sender returns a dial tone to the calling party, and then receives and temporarily stores the digits dialed. As soon as all seven digits have been recorded, the originating sender calls for an idle marker and transmits the called line number to it. In large central offices handling considerable traffic, separate markers are usually provided for dial-tone connections and for completing calls.

Upon receiving the digits of the called number, the marker recognizes from the office code that the call is for another crossbar office. The marker, therefore, seizes an idle outgoing sender, transmits to it all seven digits

Fig. 1-10. No. 1 Crossbar System

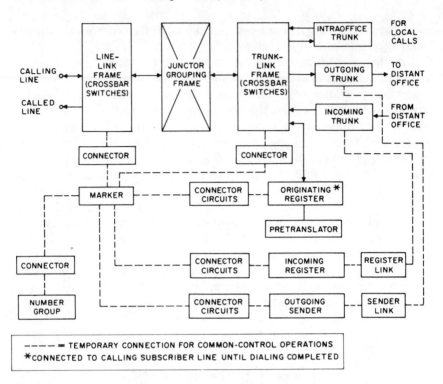

Fig. 1-11. No. 5 Crossbar System

received from the originating register, and, since an interoffice call is involved, directs the outgoing sender to prepare to transmit only the last four digits corresponding to the called line number. At the same time, the marker selects on the trunk-link frame an idle outgoing trunk circuit to the distant office, and closes a path through the junctor group frame, to connect the calling line on the line-link frame to this selected outgoing trunk on the trunk-line frame. With the completion of these functions, the marker signals the outgoing sender to prepare to pulse out the digits of the called number to the distant central office, and releases.

The incoming trunk in the distant central office is the terminating end of the previously selected outgoing trunk in the calling office. This incoming trunk is connected to the trunk-link frame, and to a register-link frame in the called office (as indicated in Fig. 1-11) which now will be assumed to represent also the called or distant office. When the incoming trunk is seized by the outgoing trunk circuit in the calling office, it selects an incoming register circuit. The incoming register will now signal the outgoing sender in the calling office to pulse out the four digits of the called line number. This information is transmitted in the form of multifrequency (MF) tones and recorded by the incoming register. An idle marker is next selected by the incoming register and the four digits of the called line number and other related data are sent to it. The marker first connects to a number group frame in order to obtain the particular line-link frame and location of the called line number, as previously explained for the No. 1 Crossbar System. Upon receiving this information, the marker connects to the indicated line-link frame and tests the called line number. If it is idle, the marker selects a path through the junctor group frame to connect the called line's location on the line-link frame with the incoming trunk on the trunk-link frame, and then releases. If the called line tests busy, the marker instructs the incoming trunk circuit to cause its associated outgoing trunk circuit in the calling office to send a busy signal to the calling party. The outgoing trunk circuit supervises the call and also controls the disconnection of the operated crosspoint contacts of the crossbar switches on the line-link and trunk-link frames when the calling party hangs up the phone.

The originating sender in the No. 5 Crossbar System is employed for some 15 to 20 seconds on a call, depending on the time needed by the calling customer to dial the called number. The marker requires only a fraction of a second to perform its functions of testing and selecting an idle trunk, connecting links, and actuating the magnets of the selected crossbar switches. A marker is connected to an originating or incoming register by a large number of wires so that the necessary data can be exchanged within a few milliseconds. Thus, a marker can quickly complete a call and be available to handle thousands of calls per hour.

Electronic Switching Systems

Direct control switching systems of the step-by-step (S $\times$ S) type comprised about 40 percent of the Bell System's estimated 18,000 central offices in 1981. And most of the approximately 13,000 dial offices operated by non-Bell telephone companies in the United States also were of this category. The nearly 11,000 common-control switching offices in the Bell System were divided among 6,500 crossbar and 4,500 electronic switching types. Electronic switching offices now make up over 40 percent of all common-control offices, whereas none was in service in 1965. The switching equipment in the electromechanical central offices is still serviceable and can be expected to perform effectively for many more years. However, the stored-program-control electronic switching equipment, now available, not only provides many new customer services but offers savings of approximately 60 percent in maintenance expenses and 40 percent in required building space. Consequently, it can be anticipated that most of the remaining electromechanical central offices will be replaced in the near future.

The need for electronic switching derived in part from the slow switching speeds of electromechanical systems, including the crossbar category, and the ever increasing demands by customers for new and improved kinds of telephone services. In electromechanical systems, the development of new and more sophisticated circuits to meet these requirements had been restrained by the inadequate operating speeds of relays and related mechanical apparatus. As a result, research and development activities turned to electronic switching, especially since this technique had been successfully employed in digital computers of various weapons systems during World War II. The invention of the transistor, subsequent improvements in solid-state devices, such as large-scale-integrated circuits, and the application of computer programming methods to processing telephone calls made possible present electronic switching systems.

The great importance of switching speeds may be illustrated by the fact that several *milliseconds* (one millisecond equals one thousandth of a second) are required in electromechanical systems for the operation of relays to perform switching and related functions. In contrast, transistors, diodes, and other electronic devices can change state or switch circuit conditions at speeds of only a few *nanoseconds* (one nanosecond is equivalent to one billionth of a second or 10^{-9} second). This ultra-high speed justifies the greater complexity of electronic switching systems. Moreover, these high switching speeds make it feasible for a single common control element, for example, the central control unit in a No. 1A ESS electronic switching office, to serve as many as 128,000 telephone customers. In contrast, the No. 5 Crossbar System can handle a maximum of only 10,000 lines, and it also requires a considerable number of duplicate groups of common control equipments.

Other important advantages of electronic switching systems include the anticipated long service life of the equipment, and its capability of adapting to new features and services without changing the system design. This design flexibility is the result of utilizing the stored program control and changeable memory concepts for processing calls in electronic switching offices. The stored program control (SPC) feature is based on logic and memory principles employed in electromechanical switching systems. Memory actually is information that is recorded and may be considered as the specific procedure to follow, or *what* to do next, in processing a telephone call. Logic makes the decision of *how* to do it. For example, memory knows what equipments in the central office should be connected to route a particular call, while logic decides what paths will be used for interconnecting the equipments. As a more pertinent example, in the No. 5 Crossbar System, the originating registers and outgoing senders provide the memory, while the markers make the logical decisions. These precepts will be further explained as we examine the applicable equipments and circuit elements of electronic switching systems in this book.

Questions

1. What is the designation of the telephone switching network serving North America? How many digits must be sent over this network to reach telephones in Canada and Puerto Rico?
2. What facilities in addition to routing and completing calls are furnished by local central offices? What is the network class designation of an end central office and a toll central office?
3. What is the maximum sequence of line numbers for a central office code in a Class 5 office? What distinguishes an area code from a central office code?
4. What are the international access dialing prefixes? What is the maximum number of digits that may be dialed after dialing the international access code?
5. Name three or more important functions performed by the telephone instrument or station subset.
6. How many pushbuttons are provided in present Touch Tone® or touch calling keysets? What particular pushbuttons are used primarily with electronic switching offices? How many audio tones are provided in pushbutton keysets? How many tones are sent for each digit?
7. What are the two general categories of switching systems? What is the main difference between them? Name two or more electromechanical switching systems still in operation in the larger cities.
8. What is the main difference between the No. 1 and No. 5 Crossbar Systems? What is the principal control device used in both systems? How long is this control device normally utilized for processing a call?
9. What type of switching system is still in use in about half of all central offices at the present time? What percentage of central offices are currently using the electronic switching type? What two major disadvantages of electromechanical switching systems instigated the development of electronic switching?
10. What is a nanosecond? Define memory and logic with respect to switching systems and give an example of each.

2
Concepts of Electronic Switching Systems

Design Principles

The development of electronic switching systems has been the result of many new concepts which are entirely different from the designs of electro-mechanical systems. The relays and crossbar switches of the No. 5 Crossbar System and the selectors and sequence switches of the panel system, for instance, are not replaced by transistors, diodes, integrated circuits, or other electronic circuitry. Instead, stored program control methods like ones used in computers, including programmed logic and temporary memory techniques, are utilized. The procedure for handling calls, whether originating or terminating, is actually divided into a series of many discrete steps. For example, the processing of an outgoing call starts with the initial handset off-hook condition, and proceeds through various operational steps until the called line is reached and the called party answers the ringing signal. The entire call-handling operation, however, appears to be accomplished almost instantaneously as far as the telephone user is concerned.

The program instructions for handling a call are stored in memory circuits. They employ the binary symbols and other elements of the machine language (designated *software*) commonly used with computers. The subsequent translation and interpretation of the program instructions are directed by wired logic in a central control unit of the electronic switching equipment. These operations generally follow the format used in computers and other information processing machines. For instance, information may be shifted from one element or register of the central control unit to another for comparison with information received from a third element. At the same time, the difference between these information inputs or received data may be stored in another register and even read out in still a different one.

The central control unit with its wired logic may be considered as the most important concept of common control in electronic switching systems. It governs the basic information processing operations, but not the switching logic which is contained in the stored program as explained later.

Types of Analog Electronic Switching Systems

Speech, tones, and other signals from telephone stations are *analog* in form. That is, they may be represented by voltage or current waveforms that continually vary in voltage or current magnitudes, corresponding to the signal's amplitude variations. *Analog* signals are transmitted and switched on a two-wire basis (for both directions of transmission) by central office equipment.

Recent advances in semiconductor technology, such as large-scale integrated circuits (LSI), which permit many complex functions to be performed by a single miniature semiconductor "chip" (integrated circuit), have made it feasible to utilize *digital* switching. *Analog* signals must be converted to *digital* form for digital switching purposes. This conversion is accomplished by sampling and time-division techniques (usually employing the 8-bit per sample method used in pulse code modulation (PCM) as described in Chapter 9). Moreover, digital switching necessitates four-wire circuits for handling both sides of a conversation on calls. Consequently, digital switching initially is being utilized for toll centers (Class 4 offices) as in the No. 4 ESS described in Chapter 9. Both electromechanical and electronic switching central offices are designed for *analog* operations.

Analog electronic switching systems may be classified as *space division* and *time division* (also designated time-division multiplex). Both employ similar techniques of line and trunk scanning, stored program control, call store memory, and central control logic functions. Different control and transmission procedures, however, are used in handling calls. In the *space division* arrangement, an individual metallic path is established through the switching equipment between each calling and called line, or trunk circuit. The *time-division multiplex (TDM)*, on the other hand, utilizes a common transmission path that is time-shared by all connecting lines and trunks through a series of high-speed electronic gates, as explained below.

During the past decade, many different kinds of analog space-division electronic switching systems have been developed and installed in numerous central offices in the United States, as well as Canada, Europe, and Japan. Table 8-1 lists some of these systems including those described in this book.

Digital Electronic Switching Developments

Electronic switching systems initially were designed to work with the other analog switching and transmission elements of the telephone plant. Most utilize the stored-program control (SPC), the concentrated permanent and temporary memory banks, and the central processing unit (central control) concepts which are described in this book. The vulnerability of central control to outages of major or significant parts of the electronic switching system required that the principal control components be duplicated. Moreover, central control could perform only one task at a time when processing calls.

These factors limit the traffic-handling and expansion capabilities of analog electronic switching offices. Furthermore, there is an increasing need for the switching network to carry both voice and non-voice (data) traffic. This requirement can be more effectively accomplished by digital switching techniques, without adversely affecting the cost or performance of the voice traffic operations.

Recent significant advances in semiconductor technology have expedited the change to digital switching systems. Inexpensive *microprocessors* and large-scale integrated (LSI) circuits have become available which can provide considerable quantities of memory capacity and greatly increase processing capabilities at lower costs. Thus, memory and processing facilities can be furnished in small units and be distributed throughout the digital switching system. This feature will permit higher processing speeds to be attained as well as improve reliability. Moreover, these technical improvements will tend to hasten the conversion from analog to digital switching, and to further improve related transmission and signaling functions for toll (Class 4) and end (Class 5) central offices.

Microprocessor technology, consequently, has made it viable and economic to utilize the *distributed control* technique in place of the central control processing currently employed in analog electronic switching systems. The distributed control method enables almost all call control functions to be handed by microprocessors. For instance, each microprocessor can control a limited portion of the digital office operations. And the microprocessor can be associated with a particular small group of subscriber lines. Therefore, a failure in the functioning of the microprocessor will harm service to only a few customers. It will not affect the central control operations of the entire digital switching office.

Principles of Time Division Multiplex Electronic Switching

The common transmission path or highway of time division electronic switching systems may be compared in some respects to time division multiplex (TDM) telephone carrier systems used to provide many voice channels over the same pairs of conductors. For example, in a TDM carrier system, such as the pulse code modulation (PCM) type, a particular voice signal is sampled on a repetitive basis and transmitted in a definite time sequence with respect to the samples of the other voice channels on the same conductors of the cable.

To assure accurate transmission and reproduction of the original speech, the voice signals must be instantaneously sampled at periodic intervals. Also, the sampling rate must be at least twice the highest significant frequency of the voice signal transmitted. In this connection, voice channels normally are designed for a frequency range of about 200 to 3300 Hertz (Hz) or a nominal bandwidth of 4000 Hz (4 kHz). Therefore, a sampling rate of 8000 Hz is usually provided so that two samples will be obtained during each half-cycle of the speech signal. These speech samples are then converted into a series of pulses for transmission over a common path, separate cable pairs being used for each direction of transmission. In a similar manner, a number of telephone calls can time-share the same switching path or transmission highway in a TDM type of electronic switching office.

Figure 2-1 is a simplified block diagram illustrating the time-shared principle used in TDM type electronic switching equipment such as the Bell System's No. 101 ESS for PBX use. Assuming that an intraoffice connection has been established between calling line A and the called party B, both lines will be periodically connected (but not simultaneously) to a common

Fig. 2-1. Major Elements of TDM Electronic Switching System

transmission path through their corresponding two-way electronic gates 1 and 2 in sequence with their assigned time slots. Whenever line A is connected in its assigned time slot (by electronic gate 1) to the common transmission path, an idle control link, LA, likewise will be connected to this common path at the same instant under guidance of the central control unit. A sample of the speech signal from line A will be stored temporarily in this link circuit. When called line B is next connected to the common transmission path in its assigned slot by electronic gate 2, control link LA again will be connected. The speech sample previously stored in control link LA will be transmitted to the called line B during its time-slot period. The same procedure will take place for the speech samples of line B for which another control link, LB, would be selected. For calls to or from other central offices, control links associated with the trunk switch units would be utilized in a similar fashion as for intraoffice calls.

The appropriate control pulses to operate the respective two-way electronic gates and the control links are sent by the central control unit, in the proper time-slot intervals. Central control receives the required data for these operations from its associated call and program memory circuits and from the line information store as indicated in Fig. 2-1. In a similar manner, other telephone lines in TDM electronic switching offices can be interconnected utilizing assigned time slots in the common transmission highway. The TDM electronic switching method is satisfactory for PBX installations and small central offices serving a few thousand lines. However, there were many difficulties in adapting the TDM process to the high-calling rate and traffic volume requirements of large central offices. These factors led to the development of the space division method of electronic switching.

Space Division Electronic Switching Principles

The space division category of electronic switching is exemplified mainly by the Bell System's Nos. 1, 2, and 3 ESS types and General Systems' Nos. 1, and 2 EAX. Although the descriptive data and explanatory information in this book pertain primarily to these types, the principles, techniques, and equipments are applicable in many respects to other space-division electronic switching systems, such as the ITT Telecomm TCS-5 and the North Electric ETS-4.

The principal elements of the space division type, as illustrated by the No. 1 ESS, may be grouped into two main parts: the central processor, and the switching network, with each part functioning separately. The central processor consists of central control, program store, and call store. The following peripheral principal elements are associated with the central processor as shown in Fig. 2-2: line, junctor, and trunk scanners, signal distributors, and the master control center.

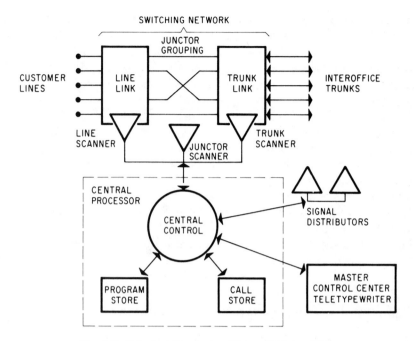

Fig. 2-2. Principal Elements of Space Division ESS

There is a continuous exchange of information between central control and its associated call store and program store. This information concerns calls in progress, particularly the status of customer lines, paths in the switching network, and outgoing trunks. Instructions for guiding the call-processing operation are contained in the semipermanent memory of program store. The call store derives its name from the fact that it temporarily stores information about a call in progress for the guidance of central control.

The switching network, comprised of the line-link and trunk-link units, interconnects calling customers' lines with selected outgoing or intra-office trunk circuits. The paths selected through the switching network consist of pairs of metallic contacts or crosspoints enclosed in special reedcapsule-type switches called *ferreeds, remreeds,* or *correeds* as explained later. These particular crosspoints, however, do not make or break circuits in which current is flowing. Relays or semiconductor devices in junctor, trunk, or service circuits provide the final closing or initial opening of the switching path to cause or interrupt current flow.

Ferrod lines, junctor scanners, and trunk scanners are used respectively to constantly monitor customer lines, paths, and trunk circuits for any changes in their status. For example, a customer's line on-hook may indicate a disconnected call, or if off-hook that a call has been initiated. A junctor or trunk circuit may be idle or busy. The conditions observed by the electronic scanners are periodically sent back to the central control unit.

The master control center administers the operation and maintenance of the electronic central office. It includes alarms, lamp displays, control keys, line and trunk test panels, an automatic message accounting (AMA) recording unit, and in the older units equipped with magnetic-core memories, a memory card writer. A data set, video display unit, and teletypewriter provide information concerning system operations and trouble reports. They are also used to type information into the switching system such as temporary translation changes in the call store memory or in operations of central control.

The foregoing presents only a very brief description of the basic elements involved in a small space division electronic switching system. There are many more components and sophisticated devices in a complete operational system such as subsequently explained for the No. 1 ESS and other electronic switching systems.

Manual and Automatic Switchboard Operations

Explanations of the principal concepts involved in the design and operation of modern electronic switching systems may be simplified by an analogy with respect to the operations of the antiquated manual switchboard. (A similar procedure was followed in discussing electromechanical switching systems in Chapter 1. References also will be made to applicable operations of the No. 5 Crossbar and other electromechanical systems as appropriate.)

First, let us visualize a multiposition manual switchboard controlled by a human operator. Normally, each operator watches the line lamps on the panels as shown in the photograph of Fig. 2-3. Each customer's line is connected to its own jack and line lamp, and has an adjacent four-digit number plate which identifies the line number. When a line lamp lights, the operator inserts the plug of an answering cord into the jack below the lamp, obtains the called line number, and completes the call in accordance with the procedure outlined in Chapter 1. Note that an operator devotes complete attention to completing one call before undertaking another one. Other customers initiating calls, therefore, have to wait their turns.

In contrast to a manually operated switchboard would be one operated electronically or automatically. An automaton, because of its almost lightning speed, can handle all positions on the manual switchboard much faster than human beings or mechanical devices. Its electronic eyes (ferrod scanners) periodically scan (about every 100 milliseconds) all line lamps on every switchboard position, one at a time. Whenever a lamp lights, our

Fig. 2-3. Manual Switchboard Position (Courtesy of AT&T Co.)

automaton stops to take appropriate action before proceeding to scan other lamps. For example, assume that line lamp 1234 lights, our electronic operator immediately consults its electronic slate (call store) concerning the previous information recorded about this particular line. Finding that line 1234 previously was idle (on-hook state), our automaton deduces that a call has been initiated, connects dial tone, and continues its scanning operations and other functions.

The digits dialed by the calling customer (for example, 555-4431) are recorded, one at a time as dialed, on the electronic slate by our automaton who also continues to perform scanning and other jobs in the interim. For instance, suppose that line lamps 7890 and 2112 lighted. Our electronic operator would enter this information on other vacant rows on its electronic slate, connect dial tone to these new calling lines, and proceed to record the called line numbers in a similar fashion. These various operations are interlaced with the work of serving other customer lines.

Returning to line 1234, when all digits of the called number (555-4431) have been recorded on the electronic slate, our automaton refers to the information stored in its semipermanent memory (program store) to obtain routing and related instructions concerning the called line number. It then proceeds to complete the call by selecting an idle outgoing trunk in the jack field above the line lamps as shown in Fig. 2-3. At the same time, it erases the information on this call that had been previously recorded on row A on the electronic slate. That row now will be available for use on another call. Figure 2-4 illustrates some of the possible entries that might appear on this fictional electronic slate.

The above analogous representations will help to explain the functions of two essential techniques of electronic switching: the single ultra-speed central control unit represented by an electronic operator or automaton, which is time-shared by all customer lines, and the temporary memory concept (call store) illustrated by the erasable electronic slate. The same manual switchboard analogy can be further utilized for explaining the functions of the semipermanent memory or program store. For instance, to handle the many diverse types of telephone calls, it would be necessary for our electronic operator to memorize what precise steps to take in processing the various classes of calls. Any changes in the established procedure, however, would necessitate retraining our automaton. A more effective and simple method is to provide a step-by-step program of instructions or an operating manual. This manual could contain simple instructions for handling calls such as, "Scan line lamp 1234. If it is lighted, write 0 in row A, square 1 on

ROW NO.	SQUARE NO.														
	CALLING LINE NO.				LINE STATUS	CALLED NO. DIALED									*
	1	2	3	4	1	2	3	4	5	6	7	8	9	10	11	12
A	1	2	3	4	1	5	5	5	4	4	3	1				1
B	4	5	7	6	0										
C															
D	7	8	9	0	1	2	0	1	5	5	5	1	2	1	2	0
E															
F	2	1	1	2	1	6	3	7	1	2					
G	4	5	7	6	1	3	1	2							
H															
I															
J	4	4	5	8	1	0									1
K															

LINE STATUS: *DIALING STATUS:
0 = ON-HOOK   0 = AREA CODE DIALED
1 = OFF-HOOK   1 = DIALING COMPLETED

Fig. 2-4. Analogous Electronic Slate

the electronic slate. If it is not lighted, write 1 in the same place on the electronic slate." Or, "Test line 7890 for busy or idle condition. If busy, refer to page 73, paragraph 37.1. If idle, refer to page 38, paragraph 16.4." In this manner, our electronic operator has to be trained only to execute the indicated steps of instructions written in the manual (program store). To change the procedure for any type of call, it is necessary only to write a new manual or program using the same type of instructions. Our electronic operator will continue to function in the prescribed manner since the knowledge for processing calls is stored in the instruction manual (program store) and not in its memory.

In the foregoing analogy to the manual switchboard, we have assumed an electronic operator (central control) using one electronic slate (call store) and an instruction manual (program store). Ideally, a single automaton is capable of working continuously 24 hours each day. To ensure continuity of operations, however, another automaton with its own electronic slate and instruction book is provided. Both automatons work together on every call, but one is placed in charge to actually control calls such as manipulating the switchboard cords and keys. The two automatons watch the same line lamps and other lamp signals, both independently make entries in their electronic slates and refer to their instruction manuals. In this way, cross-checks are made continuously between the functions accomplished by our two electronic operators. Thus, any differences or possible trouble indications in the instruction manual, electronic slate, or an automaton's operations will be immediately disclosed.

In the event of a malfunction, our electronic operators would follow the related instructions in the manual. If one electronic slate should become defective, both automatons can read from the other one although only the automaton in charge can make entries in the working electronic slate. Similarly, if the instruction manual needs repair, both electronic operators can read from the other one. Likewise, it is possible for one electronic operator to utilize both electronic slates and instruction manuals. Moreover, any trouble conditions encountered would be immediately referred by our automatons to the local test desk (master control center) in accordance with the instruction manual. Thus, maintenance personnel will be kept informed of the situation and will be able to initiate the required remedial measures.

Functions of Basic Electronic Switching Elements

Various sophisticated components and special devices have been developed for electronic switching systems. Figure 2-5 is a block diagram of such elements used in the original No. 1 ESS. Brief descriptions of these elements and their functions follow.

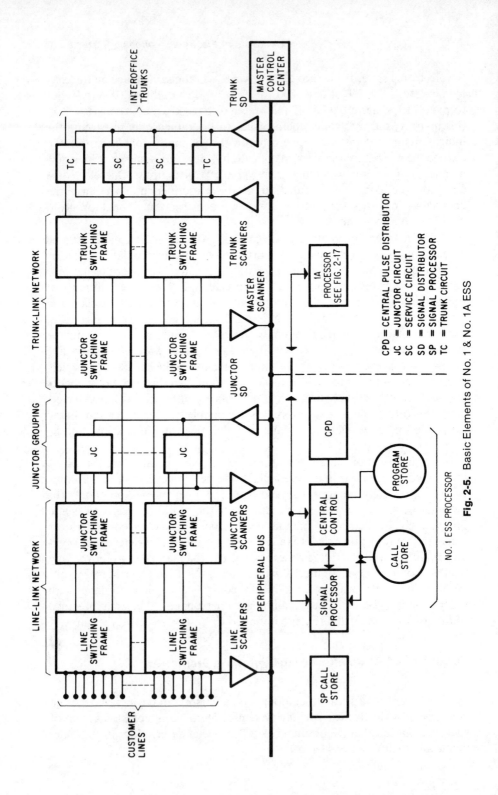

Fig. 2-5. Basic Elements of No. 1 & No. 1A ESS

CPD = CENTRAL PULSE DISTRIBUTOR
JC = JUNCTOR CIRCUIT
SC = SERVICE CIRCUIT
SD = SIGNAL DISTRIBUTOR
SP = SIGNAL PROCESSOR
TC = TRUNK CIRCUIT

INTEROFFICE TRUNKS

MASTER CONTROL CENTER

TRUNK SD

TC   SC   SC   TC

TRUNK-LINE NETWORK

TRUNK SWITCHING FRAME

TRUNK SWITCHING FRAME

JUNCTOR SWITCHING FRAME

JUNCTOR SWITCHING FRAME

TRUNK SCANNERS

MASTER SCANNER

1A PROCESSOR SEE FIG. 2-17

JUNCTOR SD

JUNCTOR GROUPING

JC   JC

LINE-LINE NETWORK

JUNCTOR SWITCHING FRAME

JUNCTOR SWITCHING FRAME

JUNCTOR SCANNERS

LINE SWITCHING FRAME

LINE SWITCHING FRAME

LINE SCANNERS

CUSTOMER LINES

PERIPHERAL BUS

CPD

PROGRAM STORE

CENTRAL CONTROL

CALL STORE

SIGNAL PROCESSOR

SP CALL STORE

NO. 1 ESS PROCESSOR

Line-Link and Trunk-Link Networks

The switching network in the No. 1 ESS is formed by the line-link and trunk-link networks with the interconnecting junctors on the junctor grouping frame. This arrangement, which corresponds to the jacks and cord equipment on the manual switchboard in the analogy of our electronic operator, is shown in Fig. 2-6. Each switch frame has a two-stage network so that the complete switching network comprises eight stages. The actual switching operations are performed by unique crosspoint ferreeds which are a form of miniature glass-enclosed reed switches. These reed switches are operated or released by controlling the magnetization of two adjacent metallic plates made of Remendur, as explained later. The switching network has a maximum capacity of 65,000 lines and 16,000 trunks.

A line-link network contains four line switch frames and four junctor switch frames. Each line switch frame includes the ferrod sensors and bipolar ferreed switches associated with the customer lines that it serves. Normally, sixty-four customer lines have access to sixteen links for a 4 to 1 concentration ratio. The particular concentration ratio employed depends upon the type of customer line. For example, business and PBX lines, because of their higher calling rates, may be arranged for a 2 to 1 concentration ratio while the service needs of residential customers may be met with a 4 to 1 ratio.

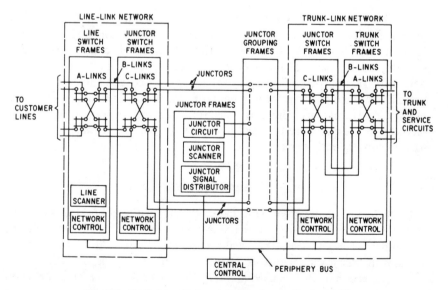

Fig. 2-6. Switching Network in Electronic Central Office

A trunk-link network usually has four junctor switch and four trunk switch frames. The number of these frames can be increased if the traffic volume requires a larger concentration ratio. For the usual 1 to 1 concentration ratio, one trunk-link network will have a capacity of 1,024 trunks, the same as the number of junctors. For a 2 to 1 ratio, the number of trunks can be increased to 2,048.

The crosspoint ferreeds on the various switch frames perform all path interconnections required for a call under direction of central control. Note that only two-wire paths, the tip (T) and ring (R) conductors which comprise the transmission circuit, are switched by the actions of the ferreeds.

Junctor and Junctor Grouping Frames

The line-link network connects through wire junctors (pairs of tip and ring conductors) to the trunk-line network. The wire junctors run through the junctor grouping frames, as shown in Fig. 2-6. In addition, junctor circuits on the junctor frame would be connected to the line-link network to provide transmission and supervision on intraoffice calls. Junctor frames also are equipped with junctor scanner and signal distributor circuits. Figure 2-7 shows a typical junctor frame with its jack and plug interconnections.

Switching Network

The switching network with its many links and paths available for handling calls requires some means to denote which paths and trunks are idle, and which are busy at any given time. In electromechanical switching systems, use is made of the S (sleeve) or C (control) leads as a memory device for this purpose. Selectors in the step-by-step and panel offices test these leads when hunting for idle paths. In the No. 5 Crossbar System, the *marker* performs this function. Electronic central offices make use of memory in the call store to select idle links, paths, and trunks in a special way, as subsequently explained.

Scanner Circuits

Scanner circuits are used to detect the state of lines and trunks. Scanners are associated with the line-link, and trunk-link switching operations and junctor frames. Scanners are also connected with trunk and service circuits. A master scanner controls the line and trunk scanning circuits.

The scanning device used is a ferrod sensor. Its design and operation are explained later. Line scanners are utilized to periodically check the conditions of customer lines to determine, for example, whether an off-

Fig. 2-7. Jack and Plug Interconnections on Junctor Frame (Courtesy of Bell Telephone Laboratories)

hook or on-hook state exists. These scanner operations are directed by central control in accordance with instructions received from program store in connection with call-processing. Junctor scanners observe the condition of junctor circuits, and trunk scanners monitor the trunk and service circuits. These various scanners may be compared, referring to the manual switchboard analogy, to the eyes of our operator.

Scanners perform their different functions in periodic intervals governed by central control. The scanpoints, or points to be scanned, are specified in the information (*address signals*) received from central control. The conditions observed at the specified scanpoints by the ferrods of the scanners are sent back to central control.

A photograph of a ferrod initially used as a line scanner is shown in Fig. 2-8, and its schematic diagram is outlined in Fig. 2-9. The ferrite rod has two control windings, an interrogate winding used to determine the state of the circuit, and the readout winding which advises central control of

Fig. 2-8. Ferrod Line Scanner (Courtesy of Bell Telephone Laboratories)

the circuit's condition. These four windings over the ferrite rod comprise a transformer. Current flowing in the two control windings determines the strength of the magnetic coupling between the interrogate and readout windings, each of which is a single turn of wire threaded through two holes in the center of the ferrite rod. When a pulse is applied to the interrogate winding while current is flowing through both control windings (off-hook or closed loop condition) no signal will be induced in the readout winding because the ferrite rod will be saturated by the resultant magnetic flux. The no-signal readout, therefore, is termed "0." If current is not flowing through the control windings (handset on-hook or open loop condition) when a pulse is applied to the interrogate winding, the ferrite rod will not be saturated. In this case, a pulse will be induced in the readout winding, and the readout is said to be "1." Consequently, whenever an interrogate pulse is applied to a ferrod, the occurrence or absense of a readout pulse denotes whether the circuit being scanned is open (readout = 1) or closed (readout = 0). An improved *ferrod,* employing Remendur (see page 199) material, is used in the No. 1A ESS to speed operations.

Signal Distributors

Signal distributors furnished in junctor and trunk frames are used to operate or release magnetic latching relays or to activate semiconductor devices in these circuits as directed by central control. In fact, signal distributors serve as buffers between the high-speed operations of central control and the lower-speed electronic circuits or electromagnetic relays. With respect to the aforementioned analogous manual switchboard, signal distributors may be considered as the hands of our operator handling the cords and keys on switchboard positions.

A signal distributor is divided into two parts, and each part comprises a controller which receives and stores information or address signals from central control. The two controllers operate independently but if one should fail, the other can take over control of the same signal distributor. In older

Fig. 2-9. Ferrod Used as Line Scanner Circuit

type No. 1 ESS, the controller connects to the specified latching relay's winding and applies the required current pulse to operate or release the relay. Approximately 20 milliseconds are required by the controller to complete its functions. A latching relay has a remanent core material which can retain sufficient magnetism to hold the relay operated after the operating pulse (− 48 volts) is received. The latching relay releases on a current pulse of opposite polarity (+ 24 volts). Thus, only one conductor is needed to operate or release each magnetic latching relay or the equivalent electronic circuitry in the No. 1A ESS.

Figure 2-10 is a simplified schematic drawing of a magnetic latching relay circuit used in older type No. 1 ESS. The relay operates on a − 48 volt pulse and releases on receipt of a + 24 volt pulse. The resistor, R, which is in parallel with the relay's winding, is momentarily disconnected by the contacts of the relay whenever it operates or releases. This results in a change in the current flowing through the latching relay's winding. The associated controller circuit in the signal distributor detects this current change and informs central control. Electronic circuits have largely replaced relays in the No. 1A ESS, thereby accelerating operations.

Central Pulse Distributors

Central pulse distributors are used to transmit high-speed pulses to control the peripheral units such as scanners, signal distributors, relays in trunk and service circuits, teletypewriters, and automatic message accounting recorders in the master control center. The functions of a central pulse distributor likewise may be compared to the hands of our operator in the operation of the analogous manual switchboard. Central pulse distributors are provided in pairs, up to eight, and the units in each pair operate independently of each other.

Information (address signals) is sent by central control to a central pulse distributor which, in turn, transmits appropriate pulses to many points in the electronic switching system over a common group of wires called the *peripheral bus*. The particular peripheral unit, such as a signal distributor, that is required to respond to this information on the common peripheral bus is selected by an enable pulse from the central pulse distributor. The time interval between consecutive orders from central control to a central pulse distributor in a minimum of 11 microseconds.

Trunk, Junctor, and Service Circuits

Trunk, junctor, and service circuits are associated with the switching network. They are utilized to complete and to supervise the paths that have been established through the switching network by central control. These

WIRESPRING RELAY
OUTPUTS

MAGNETIC
LATCHING
RELAY

ML

FROM
CENTRAL
CONTROL

THROUGH MERCURY
AND WIRESPRING
RELAY CONTACTS

R

RESISTANCE IN PARALLEL
WITH WINDING OF ML RELAY

RELAY
OPERATE
DETECTION
CIRCUIT

RELAY
RELEASE
DETECTION
CIRCUIT

− −48 V
+

+ +24 V
−

Fig. 2-10. Magnetic Latching Relay Circuit

circuits normally operate under the direction of central control through a central pulse distributor or signal distributor. Central control detects, by means of associated ferrods or scanners, any change in the trunk, junctor, or service circuits caused by related circuit operations, on-hook or off-hook actions by the customer, or a change of circuit or loop conditions made by the distant central office.

Trunk circuits only provide supervision and transmission functions. They are classified as outgoing, incoming, or two-way trunks depending on whether the local, distant office or both can connect to the particular trunk circuit. Connections between trunk and service circuits are made through the trunk-link network. The components in a trunk circuit may be reduced to only a pair of wires to permit an associated service circuit to handle pulsing and supervisory signals. In this case, called the bypass state, the transmission elements, such as the repeat coil, are switched out, thereby providing a metallic path through the switching network.

Figure 2-11 is a simplified schematic drawing of an outgoing trunk circuit that is connected through the switching network to a customer's line. The trunk scanner normally scans ferrod sensors at scanpoints A and B in the circuit. If the calling customer's line is off-hook, current will flow through the two control windings of the ferrod at scanpoint A. The trunk

SW   = SWITCHHOOK OF TELEPHONE SET
DI, D2   = DIODES
A   = FERROD SCANPOINT FOR CALLING LINE SUPERVISION
B   = FERROD SCANPOINT FOR TRUNK SUPERVISION
LI, L2   = RETARDATION COILS
INT   = INTERROGATE WINDING
READOUT   = READOUT WINDING

Fig. 2-11. Outgoing Trunk Circuit

scanner, therefore, will readout "0" for the same reasons as were discussed previously under scanner circuits. If the calling line is on-hook, no current will flow through the control windings and the trunk scanner will readout "1."

A supervisory battery is connected to the trunk conductors by the distant central office and this voltage will be applied to diodes D1 and D2 in the outgoing trunk circuit. If the trunk loop is normal (battery on the tip conductor and ground on the ring side) diode D1 will have a low resistance and current will flow through the control windings of the ferrod at scanpoint B. A high resistance will be presented by diode D2 under this battery polarity. The trunk scanner, therefore, will readout "0" at this time. When the called party answers, the equipment in the distant central office will reverse the battery polarity on the trunk conductors. Current will no longer flow through the control windings of the ferrod at scanpoint B because D1 now will have a high resistance and diode D2 a low resistance. As a result, the trunk scanner at scanpoint B will readout 1. During the conversation period, the ferrods at scanpoints A and B will be saturated by the respective loop currents and both readouts will be 0. When either the calling or called party hangs up (on-hook state), central control is informed of this change-of-state occurrence at scanpoints A and B by the trunk scanner, and disconnection will take place.

Junctor circuits are employed on intraoffice calls or connections between calling and called customer lines in the same electronic central office. In the older No. 1 ESS, the junctor circuit, which supplies a talking battery to each line, also contains two magnetic latching cut-through relays, CT-1 and CT-2, as shown in the simplified schematic diagram of Fig. 2-12. These relays, which are controlled by the signal distributor, are used to protect the ferreed switches in the switching network by connecting the battery potential after the switching network path has been closed, and removing the battery circuit before the network path is opened on disconnection. In the No. 1A ESS, electronic circuits replace relays in performing these functions.

Service circuits are associated with the trunk-line network for connection to customer lines or to trunks as required. They may be considered as

SW	= TELEPHONE SWITCHHOOK	A	= FERROD SCANPOINT TO SUPERVISE CALLING LINE
CTI, CT2	= CUT-THROUGH RELAYS	B	= FERROD SCANPOINT TO SUPERVISE CALLED LINE
INT	= INTERROGATE WINDING	LI, L2	= INDUCTORS OR RETARDATION COILS
READOUT	= READOUT WINDING	L3	= CROSSPOINT PROTECTIVE NETWORK

Fig. 2-12. Junctor Circuit for Intraoffice Calls

accessory circuits to perform functions which can be handled more efficiently by a few special circuits than by providing additional equipment in each trunk circuit. Some examples of service circuits are: the customer dial pulse receiver circuit which supplies the dial tone to the calling line and detects the dialed digits; ringing circuits to ring the called line, and to disconnect the ringing current when the called party answers; and multifrequency (MF) receivers to connect to an incoming trunk circuit in order to receive MF signals from the distant central office, and to convert these tone signals into a form that the particular scanner can observe in order to inform central control.

Central Control

The intelligence for controlling the operations of an electronic switching central office is in the central processor, a high-speed computer and data-processing unit. It may be compared to the brain of our operator in the manual switchboard analogy. The three basic components of the central processor in the original No. 1 ESS were central control, program store, and call store as illustrated in Fig. 2-2.

Central control is a computer-type binary digital mechanism that conducts complex logic operations in accordance with instructions in its semipermanent program store and temporary call store memories. The processing logic, such as what paths to interconnect, is contained in the many thousands of logic circuits in central control. For instance, central control decodes and executes binary encoded instructions received from its associated program store at the rate of one instruction every 5.5 *microseconds* (one millionth of a second). This rate is established by clock pulses originating in a 2 megahertz crystal oscillator circuit.

Three major classes of instructions normally are processed by central control:

1. The status of lines and trunks. For example, central control may be requested to examine the ferrod line scanners associated with customer lines for any change in condition, such as off-hook or on-hook.
2. Manipulating data received from the program store unit and temporarily depositing the results in call store for recall whenever needed.
3. Generating outputs for operating latching relays or electronic circuits in junctor, trunk, and service circuits.

In processing instructions, central control requests applicable information from its program and call stores. For example, upon receiving a request from call store for dial tone connection to a calling line, central control would ask program store for the processing instructions provided in a particular location or address in program store. The instructions sent back

Fig. 2-13. Equipment Frames of Central Control, Central Pulse Distributor, Master Scanner, and Call Stores (Courtesy of Bell Telephone Laboratories)

by program store would advise central control what to do with the information being processed and, at the same time, indicating the address or where to do it. Figure 2-13 is a photograph of central control (first four frames left to right) followed by central pulse distributor, master scanner, and call stores (last two frames).

Signal Processor

The signal processor with its associated call store is normally furnished only in the Bell System's larger size No. 1 ESS central offices. It operates simultaneously with and independently of central control. The prime purpose of the signal processor is to relieve central control of many time-consuming repetitive tasks. Thus, it serves to increase substantially the traffic-handling capacity of the No. 1 ESS central office. Depending on the size of the central office, up to two pairs of duplicate signal processors may be installed.

Just as with central control, the signal processor is a stored program-type of data processing machine utilizing computer technology. It operates under the guidance of program instructions stored in its call store unit, and performs both supervisory and nonsupervisory tasks. The supervisory functions comprise the periodic scanning of line ferrods to detect originating calls, junctor circuits to detect disconnection of calls, and trunk circuits for connection and disconnection signals.

The nonsupervisory tasks of the signal processor include the reception of signals from and the transmission of signals to other central offices, receiving Touch Tone® signals or dial pulses from calling customers' lines, detection of permanent signals on customers' lines, and the execution of program orders from call store. The signal processor accumulates in its call store information that is to be further processed by central control. Moreover, central control can transfer information to the signal processor's call store whenever the services of the signal processor are required. The temporary memory in the signal processor is similar to that of central control except that the signal processor employs its temporary memory both as a program store and a call store function.

Program Store

The program store is a semipermanent memory device that contains specific instructions governing the step-by-step operations of central control. It may be likened to the instruction manual used by the operator in our switchboard analogy. Figure 2-14 is a photograph of the original magnetic-core type program store modules in the No. 1 ESS electronic switching office. Semiconductor devices are now used as later explained.

The data contained in program store includes information for translating a customer's line number into an equipment location on the line-link frame and vice versa, equipment locations of trunks on the trunk-link frame, class of service of various customer lines, routing and charging information, and other items concerning particular lines or trunks. The information is contained in a random-access semipermanent memory which initially employed twistor wires and tiny magnets on aluminum cards or modules.

The following is a brief description of the original magnetic-type program store. Each aluminum card has sixty-four columns of forty-five tiny spots of magnetic material called *vicalloy*. A column represents the location of a binary "word" (a series of 0 and 1 characters) which is used to represent stored information in the binary number system as later explained. The forty-five magnetic spots in each column, of which forty-four spots are uti-

Fig. 2-14. Program Store Equipment Frames (Courtesy of Bell Telephone Laboratories)

lized, correspond to binary digits or bits; a magnetized spot represents a 0 and a demagnetized spot denotes a 1. A photograph of a twistor aluminum memory card used in program store is shown in Fig. 2-15.

The memory contents of the program store can be read out only. The insertion of information, termed writing, into program store necessitates the magnetizing or demagnetizing of the individual tiny magnets in each column of an aluminum card by an external card-writer unit. Electrical malfunctions in circuits will not alter the information contents of the program store. Semiconductor memory circuits have now replaced these magnetic devices.

Call Store

The call store is an erasable or temporary memory unit that is primarily used to temporarily record various types of information pertaining

Fig. 2-15. Twistor Aluminum Memory Card Used in Program Store
(Courtesy of Bell Telephone Laboratories)

to a call in progress. It corresponds, in our switchboard analogy, to the electronic slate used by the electronic operator.

Data temporarily deposited in call store may include the digits dialed by a calling customer, idle or busy status of links in the switching network, results of maintenance or diagnostic tests, charge information to be recorded on the automatic message accounting (AMA) magnetic tapes, and the digits to be transmitted on an outgoing call. In addition, changes in translation information, such as routing calls to a particular central office, can be temporarily stored in call store until these instructions can be placed into the semipermanent memory of program store.

The basic memory device initially employed in call store consisted of ferrite sheets, each about an inch square and containing a 16 × 16 array of tiny holes. Every hole had the equivalent of a copper wire threaded through it thereby providing a square-loop memory core. Semiconductor devices also have replaced these magnetic-type units. The information temporarily stored in call store is arranged in 8,192 words of 24 bits each, which can provide a total storage capacity of 196,608 binary digits (bits). The location of every word in call store is singularly identified by an address as subsequently explained.

Instructions or inputs received from central control specify whether interrogating (reading) or inserting information (writing) operations are to be performed, and the address of the particular location involved in call store. In the case of writing operations, the particular information or word to be stored (written) also is transmitted by central control.

Master Control Center

The master control center is the operational, maintenance, and administration core of the electronic switching central office. It serves as the interface between the electronic switching equipment and the maintenance personnel. The master control center has facilities for testing customer lines and trunks, alarms to indicate malfunctions, the means for system testing and control of operations, magnetic tapes for recording automatic message accounting (AMA) data for billing purposes, and special equipment for writing information on the memory cards used in the older magnetic-type program store.

Five separate parts comprise the master control center. The following is a brief summary of their functions:

1. *Alarm, display, and control circuits* that provide continuous indications of system operations, and permit maintenance personnel to control operations of the central office.
2. *Trunk and line test facilities* for maintenance of trunks, lines, and service circuits, including transmission tests, handling permanent signals on customer lines, make-busy circuits.
3. *Teletypewriter and dataspeed set channels* for communications between the electronic switching equipment and maintenance personnel, and for traffic measurements, line assignments, etc.
4. *AMA recording charges* on customer calls by providing magnetic tape records of all data pertaining to such calls. The magnetic tape is subsequently processed in an electronic data processing center. A separate frame is used for recording AMA in the No. 1A ESS.
5. *Program store card writer* for periodically updating translation information in magnetic-type program stores of older No. 1 ESS.

Figure 2-16 shows a maintenance or master control center in a No. 1 ESS with the aforementioned equipment items. The automatic message accounting (AMA) magnetic tapes are in the first frame on the left, the teletypewriter and related apparatus are in the second frame, the alarms, control, trunk, and line testing facilities are on the other three frames.

1A Processor System

The number of telephone calls that can be handled in a busy-hour period depends primarily on the processing speed and the capacity of the

Fig. 2-16. Maintenance and Master Control Center used in first No. 1 ESS (Courtesy of Bell Telephone Laboratories)

switching network in an electronic central office. This call-processing and instruction-execution speed, likewise, determines the maximum customer lines that can be served. These factors, as well as the continuing growth in telephone stations, traffic, and services during the past decade, resulted in the development of the *1A processor system* by the Bell Telephone Laboratories.

The 1A processor provides stored-program control for both local (Class 5) and toll (Class 4) electronic switching offices, such as the No. 1A ESS and No. 4 ESS, respectively. Employing the latest semiconductor technology, it can execute instructions in 700 nanoseconds (or 700 billionths of a second) and has a memory-access speed of 1400 ns. At these ultra-high speeds, over 1.4 million instructions per second can be accomplished as compared to the 180,000 per second executed by the processor in the No. 1 ESS.

Moreover, the 1A processor can be retrofitted into existing No. 1 ESS offices to upgrade them to the *No. 1A ESS* type, with some additions to their line and trunk networks. Thus, the No. 1A ESS, with its 1A processor, can handle up to 240,000 calls per second and serve approximately 128,000 lines depending on traffic conditions. Consequentally, fewer No. 1A ESS units would be needed for future Bell System growth and replacement of electromechanical offices than with the No. 1 ESS equipment.

The generic design of the *1A processor* is shown in Fig. 2-17. It is generally similar to the processor in the No. 1 ESS, as illustrated by Fig. 2-5, except for the addition of the direct memory-access innovation. This technological feature includes the *auxiliary unit bus* that connects central control to the *data unit selector* with its associated *tape controller and tape units,* and to the *disk controller and disk files* (file stores) equipment. These improvements provide, among other things, greater memory capacities in program store and call store components, faster execution of instructions, and the capability of quickly storing mass memory in the disk files to reduce processing time and improve reliability of operations.

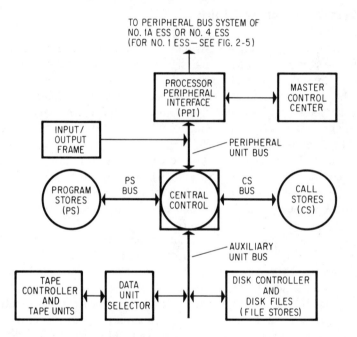

Fig. 2-17. 1A Processor

Another feature is the *input/output (I/O)* frame, which is connected to the *peripheral unit bus,* as indicated in Fig. 2-17. It serves as an interface between the relatively slow-speed terminal devices, such as dataspeed sets and teletypewriters that link master control center with central control. The I/O unit selectors and controllers furnish facilities for the operation and testing of the switching equipment by master control center, for gathering traffic data, and for making changes in data that were placed in program or call stores.

Redundancy is provided for the major elements in the 1A processor to ensure utmost service reliability. For instance, two additional program stores to the number usually needed are installed. Therefore, in the event of a failure, a spare program store can be filled with the correct data, from disk files, and substituted for the faulty one. Call stores that contain call-related temporary data are also duplicated. Likewise, two *central control* units are furnished, each executing the same tasks independently and comparing results. If an error is detected or one unit fails, the other one immediately takes over. The functions of the 1A processor and its major components in the 1A ESS are further described in related sections of subsequent chapters.

System Operation Reliability

Electronic components, switching equipment, and related circuitry in electronic switching systems are designed for continuous 24-hour operation. Thus, many years of service normally can be anticipated without interruptions because of equipment malfunctions or component failures. To ensure utmost reliability of operations and to provide for necessary maintenance and diagnostic testing requirements, the major common control elements are fully or partially duplicated.

The fully duplicated units are central control, program store, call store, central pulse distributor, and, when equipped, the signal processor with its own call store. Partially duplicated elements are the line, junctor, and trunk scanners, the junctor and trunk signal distributors, and the control circuits of the switching network.

Two central control units actually process the data on every call simultaneously and match circuits are used to compare the results at key points in the system. In case of an error or fault, central control usually repeats the operation on which the failure occurred. A continuing fault will cause central control to attempt an alternative routine or processing operation. For instance, if an error appears to originate from program store, central control may utilize a special program stored in call store which is designed to disclose malfunctions in program store. If these procedures are unsuccessful, central control will try other means, such as utilizing different combinations of duplicated units and peripheral buses, to make the system operational. Related alarm signals are generated at the maintenance or master control center to immediately alert maintenance personnel of these malfunctions.

Customer Special Services

Another outstanding feature of electronic switching systems is the capability of providing special optional services for customers by means of instructions stored in the memory of program store. The following are brief descriptions of some of the special services that are available in electronic switching offices. To provide similar optional features (other than **PBX** hunting) in electromechanical central offices would require installing special apparatus including electronic equipment, and making extensive wiring changes in the existing common control and switching equipments of these central offices.

PBX Hunting

In S × S and panel type electromechanical offices, PBX lines (Private Branch Exchange installed on a customer's premises and served by several central office lines) must have a consecutive sequence of directory numbers, for example, 5400, 5401, 5402, 5403, 5404, and 5405. This numbering sequence is necessary in order that the S × S connector switch or the final selector in the panel office can hunt for an idle line in the particular **PBX** line group, in case the first or next several lines are busy. A call to the first line in the group, 5400, can hunt through the entire group if necessary but a call to 5402, for instance, can hunt only to higher numbers and cannot hunt back to lower ones. In the No. 5 Crossbar System, it is not necessary that all PBX lines have consecutive directory numbers as long as they are in the same number group frame. A total of 1,000 consecutive directory numbers, 0000–0999, 1000–1999, 2000–2999, etc. are served by one number group frame.

The above inherent limitations of **PBX** hunting are not imposed on electronic switching systems. The number of lines in a PBX group and the order of hunting are stored in the memory of program store, and are not dependent on circuit wiring. An incoming call can be made to hunt in any desired sequence through one or more groups of PBX lines. If all lines are busy in the first PBX group, the call can be transferred to another group of PBX lines for subsequent hunting.

Class of Service

Increased versatility is also provided by the stored memory in electronic switching offices for the various classes of service and features available to customers as compared to electromechanical central offices. Approximately 100 classes of service are provided by the wired equipment in the No. 5 Crossbar System, while a total of 1,024 different service classifications can be stored in the program store memory. This large number of service classes permits the assignment of various optional services to customers as may be requested. Each optional or special feature can be assigned a specific class of service indication which could be allowed on some customer lines and denied to others.

For instance, certain PBX lines can be restricted to making calls only within the PBX, others to only local calls, some to toll calls within a particular area or to specified central offices within an area, and certain other PBX lines may not be limited in any manner. These operational functions are directed by central control when it is informed by the translation of the customer's line equipment number in program store concerning what special services the customer is entitled to.

Abbreviated Dialing

Abbreviated dialing is a special service furnished by electronic switching offices. It enables a customer to reach frequently called numbers by dialing only a two- or three-digit code instead of the usual seven, ten, or more digits. The abbreviated code is not a contraction of the called number but comprises the two-digit prefix "11" (digit * when pushbutton dialing is furnished) and another digit. The 11 or * prefix informs central control that the customer desires to use the abbreviated dial service and that only one or two more digits should be expected. A list of telephone numbers, up to a maximum of 32, which the customer may call by abbreviated dialing, is stored in the memory of program store. Customers desiring this service are assigned a one-digit number for up to ten frequently called numbers, and a two-digit number for eleven to thirty-two frequently called numbers.

Figure 2-18 depicts the steps involved in the abbreviated dialing operations. In this example, it is assumed that the calling customer dials the abbreviated dialing code *7 (corresponding to 117) which is temporarily deposited in call store by central control. Since 117 is recognized by central control as an abbreviated dialing code, this code and the calling customer's line number 7890 and its equipment location numbers are referred to program store for necessary translation. Program store finds that line number 7890 has a list of nine frequently called numbers and that code 117 corresponds to directory number 312-555-1212, and advises central control accordingly. Central control transfers this number (312-555-1212) to call store where it replaces the original dialed code 117. Central control then proceeds to process and complete the call to number 312-555-1212 just as if that number had been initially dialed by customer 7890.

Automatic Call Transfer

A customer who will be away from his home or business telephone for the day or evening can have all incoming calls automatically rerouted to another telephone number in the local area. To initiate this service, the cus-

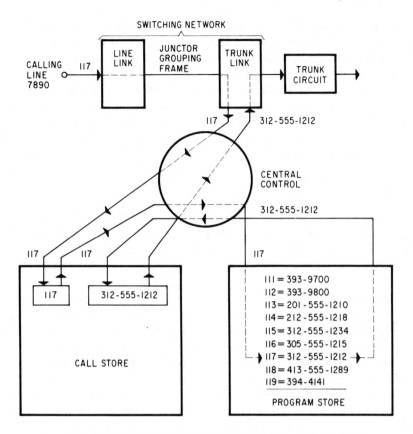

Fig. 2-18. Simplified Diagram of Abbreviated Dialing Operations

tomer first dials the assigned two- or three-digit activate code for this particular service. Upon receiving a dial tone again, he dials the directory number of the distant telephone line to which calls are to be transferred. A confirming tone will be heard as an indication that the call transfer service is activated and that all incoming calls will be rerouted to the other telephone number. The call rerouting action is directed by central control in a manner similar to that explained for the abbreviated dialing service and illustrated in Fig. 2-18.

It still will be possible, during the period of this incoming call transfer service, to originate calls from the customer's home telephone whether or not the distant telephone is in use. When the customer returns home or to the office, he can discontinue the automatic call transfer service by dialing the assigned deactivate code and receiving the confirmation tone. Subsequent incoming calls now will connect only to his own telephone line.

A preset automatic transfer of incoming calls also may be utilized. This optional service permits a customer to have all incoming calls automatically transferred to any one of eight other previously selected telephone numbers. These selected numbers are inserted in the memory of program store for the particular customer desiring this service. Consequently, any additions, changes, or deletions in the list of selected numbers must be recorded in program store by personnel at the master control center.

To inaugurate this service, the customer first dials the aforementioned activate code and when the dial tone again is heard, dials one digit to designate to which of eight telephone numbers all incoming calls should be transferred. A confirmation tone is sent to the customer as an indication that the preset transfer service is in effect. The transfer of incoming calls is effected by central control in a similar manner as previously explained for the abbreviated dialing case. To discontinue this service, the customer dials the assigned deactivate code, hears the confirmation tone, and then hangs up.

Add-On Conference Call

Another important service provided by electronic switching offices is the add-on conference call. A customer in conversation with another party can add on a third party to make a three-way conference call. For example, assume that customers A and B are conversing and that A desires to add customer C to the connection. Customer A first flashes the switchhook (equivalent to dialing 1). This momentarily releases the established connection but customer B is still held connected by his associated trunk circuit. Central control recognizes the momentary on-hook state or dial 1 digit as a request for special service, and connects a digit receiver and dial tone to the line of customer A.

Customer A next dials the add-on digit 2 followed by the directory number or (if he has abbreviated dialing service) the abbreviated code of customer C. Central control then proceeds to connect the switching network paths of lines A, B, and C to an idle conference circuit. Both A and B will hear audible ring or a busy tone depending on the state of line C. When customer C answers, all three parties will converse through the facilities of the conference circuit as shown in the block diagram of Fig. 2-19. The dotted line between the line-link and trunk-link networks in this figure indicates the original talking connection between customers A and B before connection to the conference circuit.

The add-on party, such as customer C, can be removed from the connection at any time by customer A. For this purpose, customer A first flashes his switchhook (dials 1) as in the initiation of the add-on conference service, and upon receiving dial tone dials the cancel digit 3. Central control, therefore, will release the conference circuit and will establish another talking path in the switching network between customers A and B. However, if the talking paths are busy due to traffic conditions, central control will direct that customers A and B remain connected through the previously established conference circuit.

Another party may be added any time by customer A as previously described. The conference connection is released when customer A disconnects. The usual automatic message account (AMA) records are kept for the use of the conference circuit and the calls involved. Moreover, only one toll connection may be connected to a conference circuit.

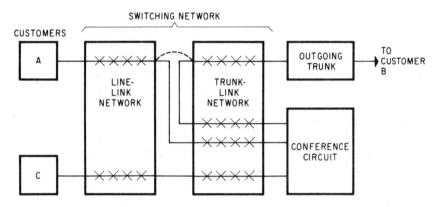

Fig. 2-19. Add-on Conference Connections to Switching Network

Call Waiting Signal

The call waiting feature, often provided in PBX systems, also can be furnished to customers served by electronic switching central offices. With this feature, a customer using his telephone will hear two short spurts of tone, about 10 seconds apart, when someone else dials his line number. This tone signal is only heard by the aforementioned customer. The calling party only hears the regular ringing signal.

The customer, upon hearing this call waiting signal, can end his conversation with the initial party by hanging up. In this event, his telephone will immediately ring and the second calling party will be connected to the line. Or, the customer can hold the initial call by depressing and immediately releasing the switchhook. This action will cause the initial party to be placed on "hold," and the customer will be connected to the other calling party. Similarly, the customer can alternate talking with either party by first depressing and immediately releasing the switchhook.

When the customer completes his conversation with either one of the two parties connected, and that party disconnects, the customer will be reconnected to the remaining party by depressing and immediately releasing the switchhook. If a third party should dial the customer's line number while the above two calls are connected to the customer's line, the third calling party will receive the normal busy signal.

Dial Conference Service

A customer can also initiate his own conference calls without the assistance of an operator by means of the dial conference feature provided in electronic central offices.This service, which is limited to one originating customer and three called parties, is initiated when the calling customer A dials the dial conference access code of two digits. This code tells central control that a conference call is to be established for customer A. Accordingly, a dial tone is sent to customer A who dials the add-on digit 2 followed by the directory number of customer B, the first desired party. Central control proceeds to establish a connection between an idle conference circuit and the lines of customers A and B as previously described and illustrated in Fig. 2-18. When customer B answers, he is informed by customer A of the conference call. Customer A now calls the next party C by first flashing his switchhook and, upon receiving a dial tone, dialing the add-on digit 2 and the directory number of customer C. When customer C answers, A repeats the same procedure to add customer D to the conference circuit.

The last party connected to the conference circuit can be removed by the orginating customer A flashing his switchhook and when a dial tone is received, dialing the cancel digit 3. All three parties are released when the originating customer A disconnects. Only one toll connection may be made to the conference circuit. The AMA equipment will record the conference circuit usage time and related data for billing purposes in the usual way.

Administration and Maintenance Features

Supplementing the special customer services that have been described are several unusual administration and maintenance features that are incorporated in electronic switching systems. These capabilities are in addition to the usual testing and maintenance facilities that are also provided in electromechanical offices, such as traffic registers, service observation, local or central test desk connections, station ringer and Touch Tone® set testing. The major items are briefly explained below.

Call Tracing

All incoming calls to a particular customer's line may be traced automatically by incorporating this requirement in the program store memory pertaining to the desired line number. If the originating line number is in the same electronic switching office as the called line that is programmed for call tracing, the teletypewriter at the master control center will print out the originating line number. If the originating line is in another central office, the teletypewriter will print out the equipment number of the associated incoming trunk.

To provide this call tracing feature, it is first necessary that the directory number translator in program store be arranged to require that all incoming calls to the specific line be traced. It is also possible to trace an intraoffice call in progress to a particular customer's line by requesting such action using the teletypewriter at the maintenance or master control center. In this case, central control will be able to determine the originating line number or the incoming trunk designation from the path memory data in call store.

Line Load Control

In electronic switching offices, customer lines are divided into two general classifications for line load control purposes, namely, essential and nonessential lines. In case of very heavy traffic loads because of storms or other contingencies, a line load control key would be operated at the master control center to give priority service to the essential customers such as police, fire, utilities, or governmental agencies. Other customers will have telephone service to whatever extent is possible, depending upon traffic conditions at any given moment.

Automatic Overload Control

If excessive delays are encountered on either originating or incoming traffic, such as connecting to idle dial pulse receivers or incoming multifrequency (MF) tone receivers, lamp and alarm signals at the maintenance control center will alert personnel. At the same time, the teletypewriter will print out a message describing the overload condition, and these printouts will continue as long as the overload lasts.

Toll Network Protection

The toll network protection feature restricts the use of specified toll trunk groups to toll essential customers. Other customers' calls destined for these toll trunks would be connected to the overflow tone. This protection arrangement is activated by the operation of a special key at the master control center which also activates the line load control operations whether or not that control key is operated. Certain customers that are classified as essential for line load control purposes likewise may be designated as toll essential customers for the toll network protection feature.

Measurements of Dial Tone Delays

Measuring dial tone delay is an important means for determining the grade of service rendered by a central office. This measurement in electronic switching offices is made by periodic test calls that are automatically initiated. The activation of line load control during overload periods also will start these dial tone delay measurements.

The number of test calls encountering excessive dial tone delays are printed out by the Teletypewriter at master control center at periodic intervals. These measurements may be started or stopped at any time by a teletypewriter request.

Service Order Changes

The temporary memory in call store is also used to record service order changes such as adding or disconnecting customer lines and trunk circuits. A teletypewriter at the business office of the telephone company may be used to transmit the service order number and associated data to the electronic switching central office. The information is recorded in the recent change register in the call store unit. A service order change is made effective when the appropriate code is dialed by operating personnel followed by the specific service order number.

Questions

1. What are the two main types of analog electronic switching systems? Which type is used in telephone central offices? Why?
2. What are the designations of the principal electronic switching systems currently developed for city and suburban central office installations? What are the two main parts of these systems? Name three or more major elements associated with the central control unit.
3. What two essential techniques of electronic switching are explained in the analogy to a manual switchboard operated by an electronic operator or automaton? Where is the knowledge stored for the manner of handling calls? What equipment unit does the electronic slate compare to in the electronic switching office?
4. What elements comprise the switching network in an electronic switching office? How many stages involve the complete switching network? What apparatus performs the actual switching operation and how many conductors are switched?

5. What are the functions of trunk, junctor, and service circuits, and with what elements of the electronic switching system are they related? What special type of relay is employed in junctor circuits? Why?
6. What is the main difference between the 1A processor in the No. 1A ESS and the processor in the No. 1 ESS? What major components are added in the 1A processor? What important element, used with central control in larger No. 1 ESS offices, is not provided in the 1A processor?
7. Define the program store and call store units. How is data or information stored in program store? In call store?
8. What main function is performed by, and how many separate parts comprise the master control center? Name at least three of these parts.
9. What major elements of the electronic switching system are fully duplicated? Name two or more elements which are partially duplicated for system reliability purposes.
10. What three or more special customer services are provided by electronic switching systems that are not available in electromechanical central offices? Name three or more unusual administration and maintenance features incorporated in electronic switching offices.

3
Central Processor
Fundamentals

Elements of Central Processor

In the previous chapter, the basic organization of electronic switching systems was discussed, including the two essential parts, the switching network and the central processor units. In the No. 1 ESS, three main elements comprise the central processor: central control, program store, and call store. In the No. 1A ESS, No. 4 ESS, and other later type electronic switching offices, these functions are performed by the 1A processor that is depicted in Fig. 2-17. They govern all system operations in handling telephone calls. In this chapter, the main fundamental functions of the central processor unit, especially central control, will be explained.

Central control may be considered as an electronic information-processing or computer-type device which serves as the center of all call-processing operations. It can request processing instructions from program or call stores, as required for handling particular calls. These requests to program store are sent as an *address* or combination of binary digits which define a specific location in program store's memory. In reply, program store will send back a two-part instruction that had been stored at the particular address. The first part tells central control what actions are to be performed, and the second part, which is the address, indicates where to do it. Central control in the No. 1 ESS decodes and executes the received instructions at a rate of one every 5.5 microseconds (5500 ns). This time period is determined by clock pulses that originate in a 2 MHz crystal-controlled oscillator circuit. Figure 3-1 illustrates the fundamental relations between the three elements of the central processor and other components of the electronic switching system. In the 1A processor, central control is designed to execute instructions on a 700-ns or 1,400-ns cycle, with an average executive rate of

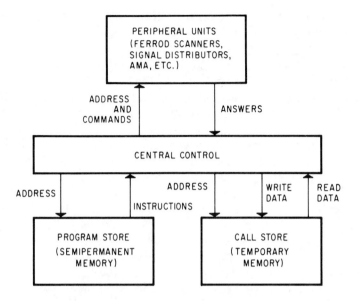

Fig. 3-1. Central Processor Elements and Interconnections

about 750,000 instructions per second. This timing originates from a 20-MHz crystal oscillator circuit in the active central control.

Three major classes of instructions are usually sent to central control from program store. The first consists of orders to sense the condition or state of lines and trunks—for instance, to check the ferrod line scanners associated with customer lines in order to detect requests for service as indicated by a change of state. In this respect, central control would perceive any change in the state of a customer's line as a binary 0 or binary 1 digit signifying, respectively, off-hook or on-hook conditions. These scanning operations are used to detect inputs to the central processor for necessary actions.

A second class of instructions concerns the processing of the input data. In this connection, central control may manipulate the received data and deposit the results temporarily in call store for later retrieval as may be needed. Additional data also may be obtained from program store as required. These operations do not progress in sequence from one step to the next. If certain conditions are encountered during any stage of the process, central control may decide to transfer to another operation sequence instead of continuing with the particular program.

The third class of instructions pertains to generating outputs from central control to operate relays or electronic circuits in junctor, service, and trunk circuits, to close ferreed switches in line-link, trunk-link, or other network paths, etc. A single instruction usually controls one operation. However, individual instructions can be combined in different desired ways for implementing any combination of control requirements.

The information and instructions deposited in program store's memory are arranged in binary words. In the No. 1 ESS, each word consists of a combination of 44 binary digits or bits. A binary digit can be either 0 or 1. Every binary word occupies a location in the program store's semipermanent memory that is uniquely identified by an address. Thus, when a particular address is received from central control, program store will send back (read out) the corresponding binary word to central control. The memory in program store is termed semipermanent because it can be changed only by external means.

Information is deposited in call store of the No. 1 ESS in a similar manner except that the binary words used contain only 24 binary digits or bits. Inputs or commands from central control to call store specify, in addition to the address or word location, the type of operation to be performed. For example, to interrogate (read) or determine the state of a particular circuit, to delete previously written data, or to insert (write) information. In the case of an interrogation (read) request, call store will transmit (read out) to central control the binary word at the indicated address. Call store contains a temporary memory because it can be changed by central control at any time. The program and call stores of the 1A processor are each capable of storing 65,536 binary words of instructions. Each binary word comprises 26 bits, which consist of 24 data bits plus 2 parity bits used for error checking purposes.

The binary digit or bit (as it is usually abbreviated) refers to characters in the binary number system. A binary digit actually represents one of two possible conditions or logic states, such as on or off, high or low potential, magnetized or demagnetized, and conducting or not conducting. These logic states are symbolized by the binary digit 0 or 1 in the binary or two-number system.

Binary Number System

The decimal system in common universal use is based on ten possible digits, 1,2,3,4,5,6,7,8,9, and 0, and the mathematical powers of 10 or 10^n. For instance, any one of the ten digits may be entered in the first space to the left of a decimal point. This digit, consequently, would be multiplied by ten every time it is moved another position to the left, for example, 7, 70, 700, 7000, 70,000, 700,000, etc.

The binary system, on the other hand, has only two digits, 0 and 1. And it operates on the mathematical powers of 2 (2^n) in contrast to the decimal system's 10^n powers. It is possible, of course, to have number systems based on other digits, such as, 3^n, 4^n, 5^n, 6^n, etc. The binary or two-digit numbering system was chosen for the basis of the machine language used by computers and similar devices because only two digits, 0 and 1, are needed to express a logic state or condition encountered, for instance, true or false, on or off, yes or no, etc.

Numbers in our decimal system may be converted to binary form and vice versa by the use of binary coded decimal tables. One form of these tables consists of columns extending from right to left with each column having a value or weight based on the powers of 2, except that the first column on the extreme right always will have a value of 1. For example, in a seven-column table, the weights reading from right to left would be: 1, $2^1 = 2$, $2^2 = 4$, $2^3 = 8$, $2^4 = 16$, $2^5 = 32$, and $2^6 = 64$. Bits that have a weight of 1, as in the extreme right columns of Tables 3-1 and 3-2, are called the *least-significant bit*. Bits in the extreme left columns of these tables, which have the highest-value weights (8 and 16, respectively), are designated the *most-significant bit*.

A 4-bit numbering arrangement used in some electronic switching elements is shown in Table 3-1. Each decimal number from 1 to 15 is represented by a binary number of four digits. Note that only binary digit 1 is assigned a weight or value, based on the powers of 2, in each of the four columns. These values start with 1 for column D at the extreme right, 2 for column C, 4 for column B, and 8 for column A. The summation of columns A + B + C + D, wherever binary digit 1 appears, will give the corresponding decimal number. For instance, decimal number 4 is composed of binary digits 0100 in that order. The decimal number equivalent, 4, is obtained

Table 3-1. Binary Number System for Decimal Digits 0–15

| Decimal Number | Four-digit Binary Code | | | |
	A (8)	B (4)	C (2)	D (1)
0	0	0	0	0
1	0	0	0	1
2	0	0	1	0
3	0	0	1	1
4	0	1	0	0
5	0	1	0	1
6	0	1	1	0
7	0	1	1	1
8	1	0	0	0
9	1	0	0	1
10	1	0	1	0
11	1	0	1	1
12	1	1	0	0
13	1	1	0	1
14	1	1	1	0
15	1	1	1	1

from the value or weight in column B in which binary digit 1 appears. In a similar fashion, binary number 0001, which has a value of 1 in column D, corresponds to decimal number 1, and binary number 0101 which has a value of 4 + 1 (per columns B and D) will be analogous to decimal number 5.

For decimal numbers between 16 and 31, it will be necessary to utilize five binary digits with values of 16, 8, 4, 2, and 1, respectively, for each binary digit 1 as shown in Table 3-2. Since the summation of 16 + 8 + 4 + 2 + 1 = 31, six binary digits or bits would be needed to represent decimal numbers from 0 to 63 inclusive, seven bits from 0 to 127 inclusive, and eight bits for expressing decimal numbers from 0 to 255 inclusive. For example, binary digits or bits 110010 represent decimal number 50, 1100100 corresponds to 100, and 11111111 to decimal number 255.

It is cumbersome, therefore, to represent large decimal numbers by the aforementioned basic binary system. Consequently, a shorthand method, based on the number 16, was devised primarily for use in microprocessors. It is called *hexadecimal* and utilizes 16 digits numbered 0 through 9 for the corresponding decimal numbers, and the decimal numbers 10 through 15 are represented by the letters A to F, respectively. To distinguish these *hexadecimal* numbers, they are usually prefixed by the dollar sign, "$." In this manner, a four-digit word or *byte* can be expressed by one *hexadecimal* digit. For instance, 1001 = $9, 1010 = $A, and 1111 = $F. Similarly, eight-bit binary words (bytes) are recorded as two *hexadecimal* digits in the range from 0 to $FF, where $FF corresponds to decimal number 255.

It is also possible to assign other binary code weights or values for the

Table 3-2. Binary Number System for Decimal Digits 16–31

Decimal Number	Five-digit Binary Code				
	(16)	(8)	(4)	(2)	(1)
16	1	0	0	0	0
17	1	0	0	0	1
18	1	0	0	1	0
19	1	0	0	1	1
20	1	0	1	0	0
21	1	0	1	0	1
22	1	0	1	1	0
23	1	0	1	1	1
24	1	1	0	0	0
25	1	1	0	0	1
26	1	1	0	1	0
27	1	1	0	1	1
28	1	1	1	0	0
29	1	1	1	0	1
30	1	1	1	1	0
31	1	1	1	1	1

decimal digits. These values do not necessarily have to follow the 2^n mathematical sequence. For instance, a four-digit binary code may have values of 5421, 6421, 7421, etc. Moreover, a 2-out-of-5 code for error detecting purposes may be utilized in certain elements of the central processor. A five-digit binary code for this purpose could have weights of 74210 for binary digit 1 in its respective five columns.

Binary Logic Principles and Circuits

The binary number system also may be considered as a form of mathematics applicable to binary logic circuits. This follows because any circuit producing on-off or equivalent type of operations possesses binary logic. And any system or subsystem having inputs and outputs which are mainly on and off operations likewise utilizes binary logic principles.

Electronic switching systems require large numbers of logic circuits, especially electronic gates, which can operate exceedingly fast and share a common control unit. An electrical circuit can be described as performing logic if it is capable of generating a high (binary 1) output only when certain specified input combinations exist. For instance, assuming two inputs, the output would be high only if the two inputs are equal, that is, both are high or both are low. The logic circuit compares the two inputs and provides a high output only when both inputs are equal.

The operation of electronic gates including the low-level logic circuits widely employed in central control may be better understood by first considering the basic types of logic circuits involved. Basic logic circuits or modules are designated AND, OR, EXCLUSIVE-OR, and NOT gates in accordance with the functions performed. Modules may be combined to form other logic circuits, such as NAND (NOT-AND), NOR (NOT-OR), and EXCLUSIVE-NOR gates. These fundamental logic functions and related circuit examples are explained below with respect to their use in electronic switching systems.

Logic AND Gate

To understand logic functions, let us consider a simple electrical circuit consisting of two switches in series that control a lamp as illustrated in Fig. 3-2. The lamp (output) will light (function) only if both switches A and B are closed to connect the power source. This particular electrical circuit is a form of a logic AND gate because both switches A *and* B must be closed for the output to function. The switches and the lamp in this circuit may be replaced with diodes, transistors, or relays to give the same AND logic functions.

The logic symbol for the AND gate is shown in Fig. 3-3. Since there are only two possible states for switches A and B, either on or off, the

Fig. 3-2. Simulated Circuit for AND Logic Function

binary number system is employed to denote the condition or state of all possible circuit combinations of switches A and B. Using positive logic, binary digit 1 is normally used to represent the "on" or "high" state and binary digit 0 the "off" or "low" state for any input or output of the logic circuit. All possible combinations of inputs to a logic circuit and the resultant outputs, such as for the AND gate in Fig. 3-3, may be presented by a tabulation of the respective binary digits in a format called a Truth Table.

The Truth Table for the simple AND gate is shown in Table 3-3. Columns A and B of this table list the binary digits representing the states of the inputs, for example, whether switches A and B in Fig. 3-2 are open or closed. Column C indicates the binary digit outputs of this AND gate. Referring to the basic AND logic circuit in Fig. 3-2, it is seen that when both switches are open, the resultant input state is represented by binary digit 0 in the first row of columns A and B of the Truth Table. Since no current flows through the lamp, the corresponding output in the table's first row of column C likewise will be binary digit 0. When only switch A is closed, column A in the second row will have a binary digit 1 but columns B and C will remain binary 0. Similarly, if only switch B is closed, column B in the third row will show a binary 1 but columns A and C will be binary 0. However, when switches A and B are both closed, current will flow to light the lamp. Therefore, the fourth or last row in this Truth Table will have binary digit 1 in columns A, B, and C.

The corresponding relations in the Truth Table also may be expressed by the following mathematical formula (with the dot meaning logic AND) called a Boolean equation:

$$C = AB \text{ or } C = A \cdot B$$

Fig. 3-3. Logic Symbol for AND Gate

Table 3-3. Truth Table for Basic AND Gate

Inputs		Outputs
A	B	C
0	0	0
1	0	0
0	1	0
1	1	1

The connective "=" denotes equivalence so that C has the same logical property as A and B.

This kind of mathematics is named after George Boole who, in 1847, introduced a mathematical form of shorthand to represent the functions of logic systems. It is not within the scope of this book to attempt to explain Boolean theorems and equations. Consequently, only the simple fundamental equations for basic logic circuits will be utilized as may be required.

Logic OR Gate

The logic OR gate, as its name implies, signifies that the output will respond to either of two separate inputs or to both. This particular logic function may be described by the lamp circuit schematic in Fig. 3-4. This schematic is a rearrangement of the switches in the AND logic circuit that was illustrated in Fig. 3-2. It will be apparent from Fig. 3-4 that the lamp will light when either switch A *or* B is closed, or if both switches are closed; hence the name logic OR gate. This type of logic circuit is also known as the INCLUSIVE-OR gate.

The usual logic symbol for the OR gate is depicted in Fig. 3-5 and its corresponding Truth Table is listed in Table 3-4. Note that the curved line between inputs A and B in Fig. 3-5 is used to indicate an OR logic function in contrast to the straight line between inputs in Fig. 3-3 for the AND logic gate.

When both switches A and B in Fig. 3-4 are open, current will not flow to light the lamp. Therefore, binary 0 digits are entered in columns A, B, and C of the first row in the corresponding Truth Table per Table 3-4. If

Fig. 3-4. Simulated Circuit for OR Logic Function

Fig. 3-5. Logic Symbol for OR Gate

Table 3-4. Truth Table for OR Gate

Inputs		Outputs
A	B	C
0	0	0
1	0	1
0	1	1
1	1	1

only switch A is closed, the second row will have a binary 1 in the A column, binary 0 in the B column (switch B is open), and a binary 1 in column C since the lamp lights. For the same reasons, if only switch B is closed, the third row of the Truth Table will indicate 011 for columns A, B, and C, respectively. When both switches A and B are closed, the Truth Table will have binary 1 digits in all columns A, B, and C as shown in the fourth or bottom row.

The corresponding Boolean equation for a logic OR gate is usually written as: $C = A + B$. This signifies that the output of an OR logic gate is equal to either A or B.

Inverter NOT Gate

The simulated logic circuits drawn in Figs. 3-2 and 3-4 may be replaced by diodes and associated components for electronic circuit applications. Diodes are passive elements and can cause attenuation to input signals. The output signal level, therefore, may not be sufficient to drive a network of additional electronic gates or other circuits. Consequently, a transistor of the common-emitter type often is connected to the outputs of the diodes in order to amplify the signals to the required level. The transistor's output signal, however, will be inverted with respect to its input signal state. That is, a binary digit 1 at the input to the transistor becomes inverted to a binary digit 0 at the output. Similarly, a binary 0 at the transistor's input changes to a binary 1 at its output.

This inversion action of the transistor results in a NOT logic gate. The logic symbol for a NOT gate is shown in Fig. 3-6. The triangle in this figure denotes the amplification action of the transistor. The small circle at the apex of the triangle indicates that the output is reversed with respect to the input signal. For example, if the input is a binary 1, the output is a binary 0. For simplicity in logic symbols, it is customary to omit the triangle and to

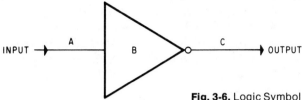

Fig. 3-6. Logic Symbol for Inventer or NOT Gate

Table 3-5. Truth Table for NOT Gate

Input	Output
A	C
1	0
0	1

represent a transistor in a logic circuit by a small circle at the output end of the particular circuit's logic symbol. This method is used in Fig. 3-7 for the NAND gate's logic symbol.

The Truth Table for the inverter or NOT gate in Table 3-5 is very simple because there are only two possible inputs and outputs. The Boolean equation for this logic function is usually written as: $C = \overline{A}$. This expression means that the output is *not* A because a bar or line over a letter signifies "not" or a reverse value in Boolean mathematical usage. For example, if $C = 1$, $\overline{C} = 0$.

Logic NAND (NOT-AND) Gate

Combining the logic AND gate of Fig. 3-3 with a transistor (inverter) is equivalent to combining the AND gate with a NOT gate as in Fig. 3-6. The result will be a logic NAND gate also known as the NOT-AND gate. Figure 3-7 is the logic symbol for the NAND gate, and its Truth Table is listed in Table 3-6.

Fig. 3-7. Logic Symbol for NAND (NDT-AND) Gate

Table 3-6. Truth Table for NAND Gate

Inputs		Outputs
A	B	C
0	0	1
1	0	1
0	1	1
1	1	0

The output of the NAND gate, per column C of its Truth Table, will be negative (binary 0) only if both inputs A and B are positive (binary 1). Note that columns A and B of the Truth Table for the AND gate (Table 3-3) correspond to columns A and B of the Truth Table for the NAND gate in Table 3-6. The outputs listed in column C of the NAND gate will be reversed as compared to the AND gate because of the transistor's inversion action.

Logic NOR (NOT-OR) Gate

The logic NOR gate, which also is known as a NOT-OR gate, is the result of combining a logic OR gate with a transistor or inverter. The usual logic symbol for a NOR gate is portrayed in Fig. 3-8. It is similar to the OR gate symbol in Fig. 3-5 with the addition of the small circle at the output to indicate the inversion action of its transistor.

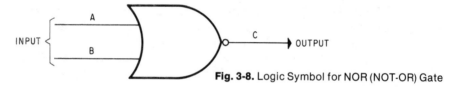

Fig. 3-8. Logic Symbol for NOR (NOT-OR) Gate

When separate transistors are required for each input, as is often the case in many electronic switching circuits employing low-level logic, the NAND gates may be used to provide NOR gate functions as in Fig. 3-9. The output of the resultant NOR gate will be positive (binary 1) only if both inputs are negative (binary 0).

The Truth Table for the NOR gate appears in Table 3-7. Note that the binary digits in columns A and B are the same as the respective columns of the OR gate's Truth Table in Table 3-4. However, the binary outputs in column C of the NOR gate's Truth Table (Table 3-7) are inverted with respect to those in column C of the OR gate's Truth Table in Table 3-4 because of the transistor.

The Boolean equation for a NOR gate may be written as:

$$C = (\overline{A + B}) \text{ or } C = (A \cdot \overline{B}) + (\overline{A} \cdot B)$$

The plus sign signifies logic OR, the dot designates logic AND, and the bar or line over the letters means "not," as previously explained.

Logic EXCLUSIVE-OR and EXCLUSIVE-NOR Gates

Logic gates OR, NAND, and AND may be combined, as illustrated in Fig. 3-14, to produce the EXCLUSIVE-OR logic circuit. Its symbol is

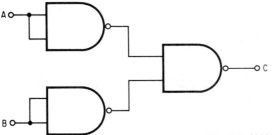

Fig. 3-9. NAND Gates Used as a NOR Gate

Table 3-7. Truth Table for NOR Gate

Inputs		Outputs
A	B	C
0	0	1
1	0	0
0	1	0
1	1	0

shown in Fig. 3-10. EXCLUSIVE-OR gates can only have two inputs. Note that a logic 1 on either input produces a logic 1 output as per Table 3-8.

The addition of a NOT gate, or transistor, to the output of the EX-CLUSIVE-OR gate changes it to the EXCLUSIVE-NOR type, as indicated in Fig. 3-11. The output of the EXCLUSIVE-NOR gate, per the binary digits in Table 3-9, will be reversed as compared to the EXCLUSIVE-OR gate.

Low-Level Logic Circuit

The basic logic gates that have been described can be merged into various circuit combinations for different electronic switching requirements. Although, for simplicity, the inputs to the logic gates in the different refer-

2 INPUTS

Fig. 3-10. Logic Symbol for EXCLUSIVE-OR Gate

Table 3-8. Truth Table for EXCLUSIVE-OR Gate

Inputs		Outputs
A	B	C
0	0	0
1	0	1
0	1	1
1	1	0

Fig. 3-11. Logic Symbol for EXCLUSIVE-NOR Gate

Table 3-9. Truth Table for EXCLUSIVE-NOR Gate

Inputs		Outputs
A	B	C
0	0	1
1	0	0
0	1	0
1	1	1

enced figures have been limited to two, many more inputs may be added as required. The same rules of logic will still prevail.

One particular logic combination, employed in central control and other electronic switching elements in the No. 1 ESS, is the low-level logic circuit illustrated in Fig. 3-12. It consists of a NAND or NOT-AND gate that incorporates a voltage limiting or bias network for the transistor. This bias circuit, indicated in Fig. 3-12 by diode D4 and resistor R2, provides large reverse bias drive currents for transistor Q1. This diode-transistor logic arrangement permits very fast switching speeds in this particular logic circuit.

In the 1A processor (used in No. 1A ESS, No. 4 ESS, and later models of the No. 2 and No. 3 ESS), improved high-speed NAND gates are used for the basic logic circuits. They employ the *diode-transistor-transistor (DT²L)* type of logic circuit as shown in simplified form by Fig. 3-13. The logic inputs in this circuit are combined and amplified by transistor Q1 and then

DI, D2, D3, D4 = DIODES
QI   = TRANSISTOR OF COMMON-EMITTER
            TYPE (INVERTER)

Fig. 3-12. Low-level Logic NAND (NOT-AND) Gate Used in No. 1 ESS

Fig. 3-13. Fast-Switching Logic NAND Gate Used in 1A Processor

inverted by transistor Q3. Transistors Q2 and Q4 function as diodes to control switching operations. These NAND gates, which were developed by the Bell Laboratories, provide ultra-fast switching with low-power requirements.

Logic Circuit Applications

The importance of logic circuits to the processing operations of central control in the No. 1 ESS processor and the 1A processor may be exemplified by the line scanning functions performed by the ferrod line scanner illustrated in Fig. 2-9. This circuit is directed by central control to scan its related customer lines at least once every 200 milliseconds in order to ascertain whether or not a change of state has occurred. The application of logic circuits utilizing NAND, OR, and AND gates for this purpose is explained in the following example:

Assume that central control is to determine whether or not a change of state has taken place for line 2345. Central control will direct call store to look in the address of line 2345 and read out its last recorded state. At the same time central control will obtain the present state of this line from the associated ferrod line scanner circuit. Referring to Fig. 2-9, recall that if line 2345 is on-hook, its ferrod sensor will produce a readout pulse, and a binary digit 1 will be returned to central control. If the line is off-hook, no pulse will be generated during the scanning process, and a binary digit 0 will be sent back to central control as an indication of the line's present state.

The last and present states of line 2345 are received simultaneously by the specified logic circuit in central control consisting of logic gates OR, NAND, and AND as illustrated in Fig. 3-14. If line 2345 is off-hook, a binary digit 0 will be transmitted by the ferrod line scanner to the A input of the OR and NAND logic gates. Similarly, if the last state of this line was record-

Fig. 3-14. EXCLUSIVE-OR Logic Circuit to Inform Central Control of Any Change of State of Customer's Line

ed as being off-hook, call store will send a binary 0 to the B input of the OR and NAND gates. Referring to the Truth Table for an OR logic gate in Table 3-4, it is noted that inputs of binary 0 will generate a binary 0 output. Thus, a binary digit 0 will be present at the A′ input of the AND gate. At the same time, the two binary 0 inputs to the NAND gate will, in accordance with the Truth Table in Table 3-6, produce a binary 1 output. Therefore, a binary 1 will be transmitted to the B′ input of the AND logic gate. The resultant output C of the AND gate will be a binary 0, in accordance with the Truth Table in Table 3-3, which will indicate to central control that there has been no change of state in line 2345.

If both the last and present states of line 2345 were "on-hook," there would be binary digits 1 on the respective A and B inputs of the OR and NAND gates as indicated in Fig. 3-14. The outputs of the OR and NAND gates, therefore, will be binary digits 1 and 0, respectively, as shown in Fig. 3-14. These binary inputs to the AND gate result in a binary 0 at its output C. This informs central control that the state of line 2345 has not changed. The arrangement of the OR, NAND, and AND gates depicted in Fig. 3-14 also represents an EXCLUSIVE-OR logic circuit.

Now suppose that the ferrod line scanner circuit indicates by a binary 0 that line 2345 is off-hook but that call store reports by a binary 1 signal that the last state was on-hook. The respective A and B inputs to the OR and NAND logic gates in this case will be binary 0 and 1, respectively. The outputs of the OR and NAND gates will both be binary 1 in accordance with their respective Truth Tables per Tables 3-4 and 3-6. The application of binary 1 digits to the A' and B' inputs of the AND gate will result in a binary 1 output in central control, thereby indicating that a change of state has occurred for line 2345. These various customer line conditions and the corresponding binary signals are tabulated in Table 3-10 for reference purposes.

Table 3-10. Customer Line Conditions and Resultant Binary Signals

Line Scanner Circuit (Present State)	Call Store (Last State)	Output Signal in Central Control	Significance
0	0	0	State not changed (Off-hook)
1	1	0	State not changed (On-hook)
0	1	1	State changed (Off-hook)
1	0	1	State changed (On-hook)

Basic Flip-Flop Memory Circuits

Groups of logic circuits may be combined to provide for temporary storage of information in the form of binary digits or bits of information. One particular form is the flip-flop arrangement that is extensively employed as a memory circuit in the central control structures of both the No. 1 ESS and the 1A processor. An electrical circuit is said to possess memory if, after being placed in a certain state by an input signal, it remains in that state after the removal of the input signal. For example, a latching relay that operates and locks up has memory because it does not release when the operating current is removed.

Two NAND or NOT-AND gates are often combined for a flip-flop circuit although two NOR or NOT-OR gates also may be utilized. The resultant flip-flop circuits are in Figs. 3-15A and B. The flip-flop nomenclature originated from the use of these circuits in computers and related equipments. The two stable states are usually called set and reset and abbreviated S and R, respectively. The flip-flop action permits the storage of information in binary form as will be explained. Flip-flop circuits remain in either state until an input signal or associated timing or clock pulses cause a change to the other state.

In central control of the No. 1 ESS, the usual flip-flop circuit consists of two NAND logic gates of the low-level type as shown in Fig. 3-12. The block diagram and logic symbol are shown in Fig. 3-11. The two inputs to this flip-flop are commonly designated S and R for "set" and "reset," and the two outputs by the letters Q and $\overline{Q}$. The bar or line over $\overline{Q}$ means "not

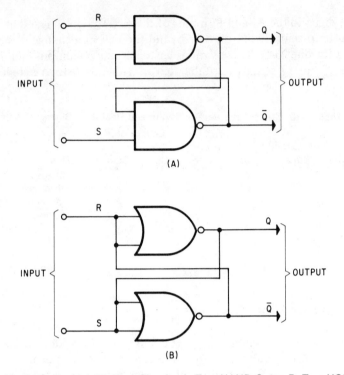

Fig. 3-15. Typical Flip-flop Circuits A: Two NAND Gates B: Two NOR Gates

Q'' as was explained for the logic NOT and NAND gates. In view of the importance of flip-flop circuits for recording information in central control and other major elements of the central processor, a brief and simplified explanation of flip-flop logic operations follows. In the 1A processor, NAND logic gates per Fig. 3-13 are provided. However, the flip-flop operations may be considered similar to those described for the No. 1 ESS processor.

Referring to Fig. 3-16, a brief synopsis of the flip-flop circuit operations is that inputs to R and S normally are in the "high" state or binary 1 whenever the flip-flop is inactive. This results because when the set (S) input is high or binary 1, the Q output will be 1 and the $\overline{Q}$ output will be 0. Conse-

Table 3-11. Basic Flip-flop Truth Table

Inputs		Outputs	
S	R	Q	$\overline{Q}$
1	0	1	0
0	1	0	1
0	0	No Change	
1	1	Undetermined	

quently, any change in the state of the R input will not affect the output of $\overline{Q}$ because the cross-connection from $\overline{Q}$ to an input of the NAND 1 gate will keep the transistor in NAND 1 nonconducting. Now, if the input to R becomes binary 0, the input to the transistor of NAND 2 will be decreased and its $\overline{Q}$ output will change from binary 0 to 1. This reversal of $\overline{Q}$ from a low to high state will increase the drive current to the transistor in NAND 1 so that it will conduct. The Q output, consequently, will change from binary 1 to 0. These actions will continue in unison with changes in the respective input states.

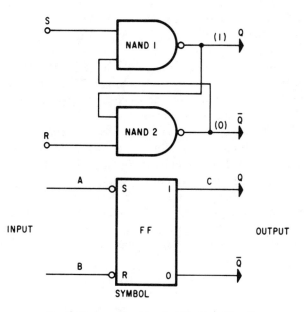

Fig. 3-16. Low-level Logic Flip-flop Circuit

J-K Flip-Flop Logic Circuit

The basic flip-flop circuit has some disadvantages especially when the inputs are simultaneously binary digits 1. This situation may be understood by examining the flip-flop Truth Table in Table 3-11. Observe that the Q output is a function of the flip-flop's outputs and its inputs at the previous time of the binary digit. For instance, if the set (S) input is binary 1 and the reset (R) input is binary 0, the Q output will be binary 1 as shown in the first row of the Truth Table. If the S and R input states are reversed, the Q and $\overline{Q}$ output states likewise will be reversed as indicated in the second row of the Truth Table by the respective binary digits. However, when the inputs are both binary 0, the output will not change from what it was in the previous bit time. Also, when the inputs are simultaneously binary digits 1, the next state cannot be determined.

To overcome this undesirable flip-flop situation and to provide for binary counting or division operations, use is made of the J-K flip-flop circuit. This device, which can handle all possible input combinations, is represented by the symbol in Fig. 3-17. This symbol comprises two basic flip-flops in series. The inputs marked J and K influence the flip-flop operations only when synchronized with the clock or timing pulses that control the operations. The set and reset inputs operate directly on the output without being synchronized with the clock pulses.

In the J-K flip-flop memory circuit, the Q and $\overline{Q}$ outputs will change to their respective opposite states when both J and K inputs are binary 1. The outputs will not change if both J and K inputs are binary 0. When a binary 0 is applied at the reset input, the output of the J-K flip-flop will be in the zero-state position or Q = 0. If a binary 0 is applied to the set input, the J-K flip-flop will be placed in the set state or Q = 1. The Truth Table in Table 3-12 illustrates these fundamental relations which are specifically applicable to the storing of information in central control.

Various combinations of flip-flop memory circuits are employed in central control and other elements of the No. 1 ESS and in the 1A processor. These include different types of registers, buffers, binary counters, and associated components. The following descriptions of the major flip-flop circuits and their operations, although predicated on the No. 1 ESS processor, are also applicable in most respects to the 1A processor system used in the No. 1A ESS, No. 4 ESS, and later types of the No. 2 and No. 3 ESS.

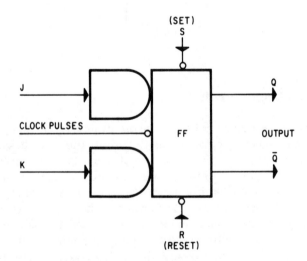

Fig. 3-17. J-K Flip-flop Symbol

Table 3-12. J-K Truth Table

Inputs		Outputs	
J	K	Q	$\overline{Q}$
1	0	1	0
0	1	0	1
0	0	No Change from Previous State	
1	1	Changes to Opposite State	

Register Circuit

The major use of flip-flops is to form a *register circuit* that can store one binary word. One flip-flop circuit is needed for each binary digit or bit comprising the stored binary word. Registers in central control of the No. 1 ESS normally have a capacity of a 23-bit binary word, which includes one bit used to represent the sign of the word. This sign bit is 0 for positive numbers and 1 for negative numbers. In the 1A processor, a binary word contains 26 bits (24 data bits plus 2 parity check bits). Also, its program store utilizes two binary words for instructions instead of the single binary word used by its call store, as later explained. Approximately 10 different registers are provided in central control for temporarily storing binary words or data during call-processing operations.

A typical register in the No. 1 ESS central control may be represented by the simplified block diagram of the NAND logic gates and flip-flop memory symbols in Fig. 3-18A. Recall from Fig. 3-12 that the output of a NAND gate will be high (+ 4 volts) when at least one of the inputs is low (0 volts), and that its output will be low only when all inputs are high. This nomenclature is used in the following explanation of the operation of the register circuit.

We will assume, referring to Fig. 3-18A, that all flip-flops in the register circuit have been reset by a momentary low (0 volts) pulse on the common reset lead that is connected to all R inputs. When a high (binary 1) is received on gate lead A, the binary information on input leads A_0 to A_{22} inclusive will be sent to the corresponding 23 flip-flops designated FF BIT 0 to BIT 22. Note that each flip-flop is marked by its related bit number to simplify the description of the operation of the register.

For example, if the input to lead A_{14} should be high, when gate A closes, flip-flop BIT 14 remains in its reset state. The same conditions apply to the other input leads in the A_0 to A_{22} sequence. In a similar manner, when gate lead B is high, the binary information on input leads B_0 to B_{22} inclusive will be forwarded to their respective flip-flops. Thus, gates A and B determine when the input data is to be sent to the register and from what source. For instance, gate A may control the inputs from program store and gate B from scanner circuits.

Fig. 3-18. Block Diagram of a 23-Bit Register and Its Symbol

The symbolic form of a register is shown in Fig. 3-18B. The control gates are often omitted for simplicity purposes. The different registers provided within central control are usually designated by the functions that they perform. For example, L for logic register, F for "first one" register, K for accumulator register, and the X, Y, and Z index registers.

Program Instruction Fundamentals

Before proceeding with the descriptions of the organization and operation of central control, it is desirable that the principal concepts of program instructions and operations associated with them be understood. The stored program in electronic switching offices utilizes a large number of various types of instructions. The program is organized so that an interrupt system initiates the input-output program instructions that must be performed in accordance with an accurate timing tolerance. Data collected by these input-output programs are forwarded to the call processing programs which decide the course of action for handling calls.

Three basic types of program instructions generally are employed. They may be identified by the functions performed, for example, the *perform type,* the *proceed type,* and the *examine type.* A perform type of instruction would request central control to carry out a specific operation and then advance directly to the next instruction. A proceed type would require central control to go unconditionally to a specified address in the program and to proceed from that point. An examine instruction would require central control first to determine the value of some specified binary quantities and then to decide whether to advance to the next instruction or to transfer the data to a specified address in the program. The decision to transfer is based on the sign and *homogeneity* of a particular binary word. In this respect, recall that one of the 23 bits comprising a binary data word represents the sign of the word, and that this sign bit is 0 for positive numbers and 1 for negative numbers. The homogeneity, which may be defined as the mathematical state of a data word, is 1 if *all* of its bits are either 1 or 0; otherwise homogeneity is 0.

The fundamental structure in the No. 1 ESS of a typical program incorporating the aforementioned types of instruction is outlined in Table 3-13. The actual program for an electronic switching office employs many instructions of various types and has a very complex pattern. However, it is similar in concept to the program shown in Table 3-13. For instance, the binary number sequence and its equivalent decimal number represent the address or location of a specific instruction in program store. Each of the letters in the horizontal blocks symbolizes a particular type of instruction. For example, in the first column, instructions A and B are of the same perform type, each advancing without restrictions to the next address. Instruction C is the examine type which would direct central control either to instruction D or to instruction Q. Instructions D to O, inclusive, also are of the perform type leading unconditionally to the next address.

Referring to column 2, instructions Q, R, and S are perform-type instructions and T is an examine type leading either to instruction U or Y. Instructions U, V, and W are all of the perform type. Instruction X is of the proceed type which leads directly to instruction L. Instructions Y, Z, AA, AB, AC, and AD in column 3 likewise are of the perform type. Instruction AE is a proceed type leading directly to instruction P in column 1. Instruction P also is a proceed type which leads back to the initial instruction A.

Table 3-13. Fundamental Structure of a Program

	1			2			3	
A*	B†	C‡	A	B	C	A	B	C
(0)	00000	A						
(1)	00001	B						
(2)	00010	C						
(3)	00011	D	(16)	10000	Q			
(4)	00100	E	(17)	10001	R			
(5)	00101	F	(18)	10010	S			
(6)	00110	G	(19)	10011	T			
(7)	00111	H						
(8)	01000	I						
(9)	01001	J						
(10)	01010	K	(20)	10100	U	(24)	11000	Y
			(21)	10101	V	(25)	11001	Z
(11)	01011	L	(22)	10110	W	(26)	11010	AA
(12)	01100	M	(23)	10111	X	(27)	11011	AB
(13)	01101	N				(28)	11100	AC
(14)	01110	O				(29)	11101	AD
						(30)	11110	AE
(15)	01111	P						

* Equivalent decimal number
† Address in binary numbers
‡ Type and text of instruction

Program Instruction Components

Program instructions, such as those received from program store, may be divided into four parts or fields. These parts are usually designated *operation* (OP), *identification constant, data address* (DA), and the *options field*. Program instructions are written mnemonically, utilizing combinations of letters and numbers as coded abbreviations for the text of an instruction.

The operation (OP) part specifies the particular work to be performed by central control, such as transferring data from an address in program store memory to the X register in central control. The identification constant is a digit which identifies the pertinent word in a block or group of words at a particular address in program store. This constant is not changed until the program itself is altered. (In the No. 1 ESS processor, this can be accomplished only by writing new permanent magnet twistor cards as later explained.) The data address (DA) field refers to the address or location of the binary data words of the specified instruction in program store or in call store. The locations in the program and call stores are part of a joint memory list in the program instruction. The binary value of the address determines whether a specified location is in call store or in program store.

The options part of the instruction is for introducing options or required variations in the execution of an instruction. For instance, the contents of the X register in central control may be moved to an address or location in call store by the complementing (C) option. In this case, each bit of the binary word to be moved is changed from 1 to 0 or vice versa. For example, the binary word 01110110 would become 10001001. Another kind of option included in many instructions is the product masking (P) option which is illustrated in Fig. 3-20 and subsequently explained.

Some segments of programs may require identical sequence of instructions at various points. To repeat the identical sequence wherever needed would involve a needless repetition and waste of program words. To preclude this situation, the particular instruction can be provided only once as a *subroutine* which is common to several programs. Figure 3-19 illustrates the operation of a translation subroutine that is common to both the translation data and dialing connection programs. It functions in the following manner: At the end of instruction C, the translation data program is transferred to the translation subroutine. The program continues through in-

Fig. 3-19. Application of a Subroutine to a Program

structions XA to XJ or through instructions XA, XB, XC, XD, to XE and then through XEA to XH, XI, and XJ. From instruction XJ at the end of the subroutine, the program returns to instruction D. A similar procedure is followed for the dialing connection program in which the transfer to the translation subroutine is made from address 202 (instruction RC). In this case, the program returns to address 203 (instruction RD). Note that at the end of a subroutine, central control returns to the appropriate point in the program.

Logic and Arithmetic Operations in Central Control

Central control processes data by manipulating arithmetic and logic quantities expressed in the binary number system which was previously explained. The arithmetic numbers contain 23 bits, designated 0 to 22, with bit 22 known as the *sign bit* because it is used to distinguish between positive and negative numbers. Addition and subtraction are the two basic arithmetic operations performed since the call processing and other system functions do not require multiplication or division. The arithmetic circuits in central control also compare, shift, and rotate binary numbers and perform certain logical operations.

The *compare operation* is related to substraction. It involves the subtraction of two numbers to determine the sign and homogeneity of the difference as a basis of a transfer decision in a subsequent instruction. The *shifting function* is often executed to transfer two items of information, found in different positions within two data words, to a position where the two items may be logically combined. For example, the dial pulse count for a decimal digit may be accumulated in bit positions 0–3 of the K register as shown in Fig. 3-20A. However, it may have to be stored in bit positions (from right to left) 4–7, 8–11, or 12–15 according to the particular digit of the called number it represents. A shifting operation is necessary, therefore, to get the accumulated data from bit positions 0-3, to positions 4–7. Another type of shift operation is illustrated in Fig. 3-20B which shows the shifting of a 16-bit word four positions to the right. Note that each bit in this binary word is moved four positions to the right so that the underlined bits, 0110, in positions 0–3 are moved out of the word and deleted. Four 0 bits are added to bit positions 12–15, which were initially occupied by the first four bits of the binary word. Shifts to a different number of positions, whether to the right or left, are handled in a similar manner.

Rotation is an arithmetic operation in which bits are passed through one end of a register in central control and reinserted at the other end but no information is lost in the process. Any number of bits or all bits in a word can be rotated as required in either direction. An example of the rotation or

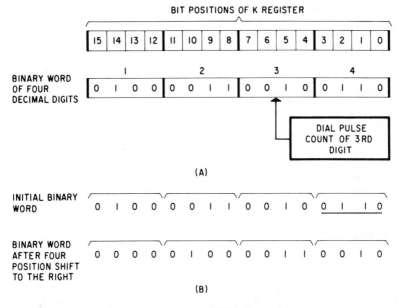

Fig. 3-20. Examples of Shift Operations

circular shift of binary numbers is as follows. The initial 16-bit binary word: 0100 0011 0010 0110. The binary word after being rotated four positions to the right: 0110 0100 0011 0010. (See Fig. 3-20B.) Note that the four-position rotation operation is similar to a shift to the right except that no bits are lost. Any bit that passes through the right end of the binary word will return through its left end. Similar procedures apply to rotations to the left. The rotation operation is used extensively in the network path hunt program.

Example of Logic Operations in Central Control

The previous described manipulations of arithmetic and logic quantities have significant relations to the call processing functions of central control. A simplified example of such applications is shown in Fig. 3-21. This illustration refers to the processing operations involved with the state of customer lines shown in Fig. 3-14 and tabulated in Table 3-10. Recall, in this connection, that the ferrods in customer line scanning circuits are scanned at least once every 100 milliseconds in accordance with the line scanning program.

In Fig. 3-21, the first line, P, represents the present or latest scanner readings in binary numbers received by central control from a particular group of sixteen-line ferrods. The 0 in bit position 0 (the first from the right end) indicates that the corresponding customer's line is off-hook. The 1 in

BIT POSITIONS

	15	14	13	12	11	10	9	8	7	6	5	4	3	2	1	0
P = PRESENT FERROD SCANNER READING	1	0	1	1	0	1	0	1	0	0	0	1	0	0	1	0
L = LAST SCANNER READING RECORDED IN CALL STORE	1	0	0	1	1	1	0	1	0	0	0	0	1	1	1	0
C = CHANGES IN STATE OF CUSTOMER LINES (EXCLUSIVE-OR LOGIC OF P AND L)	0	0	1	0	1	0	0	0	0	0	0	1	1	1	0	0
D = CUSTOMER LINES INITIATING CALLS	0	0	0	0	1	0	0	0	0	0	0	0	1	1	0	0

Fig. 3-21. Logic Operations in Customer Line Scanning

the next bit position (1) signifies that its associated customer's line is on-hook. In a similar manner, the states of the other lines in the particular scanned group are denoted by the respective binary digits 0 and 1. The second line, L, symbolizes the readouts received from a call store address in which the previous scanner readings of this same group of line ferrods had been temporarily stored. Thus, the bits in line L disclose the states of these sixteen customer lines 100 milliseconds before.

The third line, C, portrays the EXCLUSIVE-OR logic combination of lines P and L in accordance with the data in Table 3-10. The bits in line C that are 1 identify the particular line scanners in which a change of state has occurred between the present and last scanner readings. Additional logic circuits are utilized to ascertain which bit positions that have undergone a change of state (as indicated by line C) are presently binary 0, or off-hook. The resultant bit positions (2, 3, and 11) are indicated in Line D by a binary 1. The binary 1s in line D signify that the corresponding customer's line is now off-hook to originate a call. Central control, therefore, will proceed to connect to each originating line, in turn, a service circuit that will provide dial tone and detect the dial pulses for further processing.

Masking and Other Data Processing Functions

An essential data processing function is to find the required binary word in a block or group of words. To accomplish this, central control sends two inputs to the index adder circuit. One input is the identifying constant of the instruction and the other is a number representing the contents of the specified index register. The index adder adds the two and the output of the adder is the address or location of the desired binary word.

In some cases, certain bits are not conveniently located in a binary data word. For instance, the binary words contained in call store are often divided into subwords, that is, a word may have several 4-bit groups, each storing a dialed digit for a particular call in progress. In such cases, masking may be utilized to isolate a subword within the word without affecting the relative bit position. The masking operation, in effect, makes 0 all but certain specified bit positions of a word. This process is illustrated in Fig. 3-22 where the bits of binary word W are converted by a mask M so that the bits in positions where mask $M = 0$, are changed to 0. The bits in positions where $M = 1$ are not changed. The resultant word is really the logical product (AND logic function) of the word W, and the mask M in accordance with the Truth Table in Table 3-3.

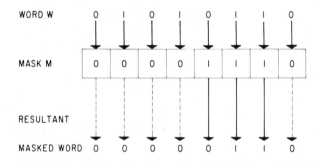

Fig. 3-22. Arithmetic Masking Operation of Word W and Mask M

Masking or the process of changing to 0 certain bit positions of a binary word, is also employed to insert information in specified bit positions of a call store address without affecting the other bit positions. This insertion masking procedure is shown in Fig. 3-23. Binary word W is inserted by mask M into address AB in call store. The procedure is that in every position where mask M has a 1, the corresponding bit from binary word W replaces the original bit of the word in location AB. In bit positions where the mask M has a 0, the corresponding bit initially in AB does not change. The resultant bits in location AB from this insert masking process are shown in the bottom row of Fig. 3-23.

Logical manipulations of binary data words follow the applicable logic functions. For instance, assume that the AND logic function per Fig. 3-3 is applied to binary words A and B. The result, as shown below, will have a binary 1 only in those bit positions where both A *and* B have a 1 (see Truth Table in Table 3-3):

				Decimal No.
Binary Word A	=	0 0 1 1 0 1 1 0	=	54
Binary Word B	=	0 0 1 0 1 0 1 1	=	43
Resultant AND	=	0 0 1 0 0 0 1 0	=	34

For the logical OR function, using the same binary words A and B, the result will have a 1 in those bit positions in which A or B both have a 1 (see Table 3-4) as indicated below:

				Decimal No.
Binary Word A	=	0 0 1 1 0 1 1 0	=	54
Binary Word B	=	0 0 1 0 1 0 1 1	=	43
Resultant EXCLUSIVE-OR	=	0 0 1 1 1 1 1 1	=	63

The contents or binary data words of a register in central control, or a word location in either call store or program store, can be moved from one location, such as A, to also appear at another location B. The contents of location A will remain unchanged in this process. Note that data can be moved from program store but not into a program store location.

Another important data processing role of central control is the right-most 1 function. This consists of detecting and identifying a binary 1 among a number of binary zeros in a data word. The binary 1 may signify the first customer line whose dial pulse count must be updated, or an idle trunk. For instance, if central control is searching for an idle trunk to complete a call, the binary zeros indicate busy trunks and the first or rightmost binary 1 signifies an idle trunk in the particular trunk group. In this operation, the specific binary word that stores the idle and busy states of the particular trunk group is first directed by central control to the K register. The right-

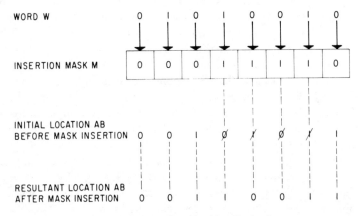

Fig. 3-23. Insertion Masking Procedure

most binary 1 is then detected by an associated detector circuit which conveys the bit position of the rightmost 1 to the F register via the masked and unmasked buses as subsequently described.

Questions

1. What are the major elements of the central processor in the No. 1 ESS? In the No. 1A ESS?
2. What is meant by an address? A word? Where are they found?
3. What digits comprise the binary number system? How does the binary number system differ from the decimal system? Are other number systems possible?
4. How may numbers in the decimal system be converted to the binary form and vice versa? What are the equivalent decimal numbers for the binary numbers 0011, 1111, 101101, and 00000101?
5. How does an electrical circuit perform logic? What are the basic logic circuits? Can they be combined to form other logic circuits? Explain.
6. What particular logic circuit is extensively employed in central control? What are its main components? What is its advantage?
7. What is a flip-flop circuit? Where is it used? How are its inputs and outputs usually designated?
8. Describe the J-K flip-flop circuits with respect to the basic flip-flop. Where are they used?
9. Explain the three basic types of program instructions. Define homogeneity as used with regard to a binary word.
10. What is the value of the following binary word after it is rotated four positions to the left?

    0100 0011 0010 0110

    What is the rightmost 1 function?

4

The Central Control Unit

Basic Structure of Central Control

The call processing functions of the central processor unit (CPU) in electronic switching systems are performed by the central control unit which, as may be expected, has a very complex structure. It can be regarded as a binary digital device that performs highly complex logic operations in accordance with stored instructions and information received from program store, and pertinent data from call store. The processing logic is contained within the many thousand logic circuits in central control which are designed to function at very high speeds. Central control directs the entire operations of the electronic switching system and, in effect, represents the highest degree of common control in telephone switching applications.

In order to simplify the explanations of central control's organization and operation, its principal elements are considered as consisting of *interface, control,* and *data processing* sections. Furthermore, a number of block diagrams are used to progressively present the major functions of central control with particular respect to the control and processing sections. Figure 4-1 shows the basic structure of central control including its interface units, and the major control and processing components used in the original No. 1 ESS. However, the following block diagrams and descriptions also apply in most ways to the No. 1A ESS and other electronic switching systems that employ the 1A processor or equivalent central processor units.

All instructions and information from program store are read out or received by the buffer order word register (BOWR). This circuit serves as an interface or isolating circuit between program store and central control. The BOWR connects to the index adder circuit and to the control section's components. The index adder receives from the BOWR the information con-

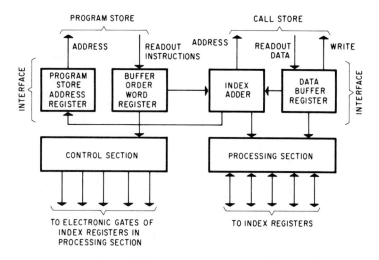

Fig. 4-1. Basic Structure of Central Control in No. 1 ESS

tained in the data-address field of the instructions transmitted by program store. The functions of the index adder will be subsequently explained. The program store address register (PAR) also interfaces with program store. It generates the addresses or locations of instructions within program store that were requested by central control through the index adder.

The data buffer register (DBR) and the index adder interface with call store. The data buffer register sends (writes) data to and obtains data (readouts) from call store. The data buffer register likewise connects with the index adder which, in addition to its other functions, generates addresses to call store whenever necessary to retrieve data that had been temporarily stored there by central control.

Control and Processing Sections

The control section directs the operation of the input and output electronic gates that are associated with the various registers in the processing section of central control. The block diagrams in Fig. 4-2 present further details of the processing and control operations. Note that input and output gates are shown with the various index registers. In subsequent diagrams, these gates will be omitted for simplicity purposes because the symbol for a register will include these electronic gates.

The order word register (OWR) with the order word decoder (OWD) are used to control, in part, the execution of instructions received by the BOWR from program store. These circuits are provided in the No. 1 ESS

Fig. 4-2. Control and Processing Sections of Central Control in No. 1 ESS

processor because of the overlap between the basic 5.5 microseconds time cycle of program store's operations, and the approximately 10 microseconds required for processing by central control. This time interval includes the period from the receipt of instructions from program store until the specified data from call store is read out into the data buffer register and processed by the processing section. As a result, there will be a time overlap in the execution of two consecutive instructions. The buffer order word decoder (BOWD) and the order word decoder, consequently, are utilized to simultaneously control the execution of the two consecutive instructions.

In addition, the BOWD and the order word decoder control the flow of processing information within central control by operating the input and output electronic gates associated with the registers in the processing section as illustrated in Fig. 4-2. This electronic control function is termed *gating,* and it will be frequently used in this book. For instance, data from call store first is gated to the data buffer register (DBR) under control of the order word decoder. The data first is conveyed over the unmasked bus, and through the masking and complement circuit to the masked bus. Each bus consists of 23 leads, one for each bit of a data binary word. The masking and complement circuit implements the options of product masking and complementing. From the masked bus, the data is sent to the specified register circuit. For instance, the X register, because its input gate was operated by the order word decoder. In a similar manner, the order word decoder would operate, for example, the output gate of the Y register. Its contents (recorded data), therefore, would be sent to call store for temporary storage via the unmasked bus, masking and complement circuit, masked bus, insertion masking circuit, and the data buffer register as shown in Fig. 4-2.

The aforementioned operations of the control section within central control may be summarized as follows: The BOWR controls the addressing of call store. The order word decoder, on a readout instruction, controls the gating of information from call store to the data buffer register and, via the unmasked bus, masking and complement circuit, and then over the masked bus to the specified index register in the processing section. Likewise, the order word decoder, on an instruction received from program store, controls the gating of data temporarily stored in an index register (such as the X register) to call store. In this case, the data from register X is sent over the unmasked bus, masking and complement circuit, masked bus, the insertion masking circuit, and through the data buffer register to call store.

Processing Binary Data Words

Each register in the processing section of central control in the No. 1 ESS can store a binary data word of up to 23 bits. The binary word may have originated in call store or in part of the instructions that were received from program store. Binary words read out from call store actually consist of 24 bits, the capacity of the DBR. The extra bit is used for parity check purposes and remains within the data buffer register.

Binary words from program store contain 44 bits as received by the BOWR. These 44 bits are divided into three groups which are called the data-address field, operation and option field, and error detection and correction circuit as shown in Fig. 4-3. The data-address field, depending upon the type of instruction, may contain bits 0 to 22. Bits 23 to 36, inclusive, normally make up the operation and option field.

The remaining bits, 37 to 43, are used by the error detection and correction circuit. The binary data words read out from program store are encoded (using the Hamming code) so that single errors can be detected and corrected. If a double error should be discovered, the particular instructions from program store will be reread. Likewise an error in an address will cause the instructions to be reread.

Referring to Fig. 4-3, bits 0 to 22 in the data-address field are gated, under the control of timing circuits and in accordance with the program instructions, to either the L register or to the index adder for further processing. The index adder circuit may modify, when required, the data-address field by adding to it the contents of a specified index register. The logic or L register is connected to the unmasked bus and the masked bus so that this register also may be controlled and read similar to any index register.

Fig. 4-3. Division by Central Control in No. 1 ESS of a 44-Bit Binary Word Readout from Program Store

The 14 bits, 23 to 36, inclusive, in the operation and option field, are dispatched to the order word decoder through the order word register. These particular bits also are sent to the BOWD as indicated in Fig. 4-3. These 14 bits in the operation and option field serve to control the gating of the several index registers in the processing section, which are illustrated in Fig. 4-2.

Masking Functions

The 23-bit word length of an index register in the processing section of central control is longer than most of the data quantities handled with call store. For instance, a single binary-coded decimal digit, such as 8, is 4 bits long. Therefore, up to five of these data quantities may be packed into one binary data word. In order to efficiently handle such partial data words, masking facilities are provided in central control for use in accordance with program instructions.

Masking functions are provided by the masking and complement circuit, the insertion masking circuit, and the masked and unmasked buses which are shown in Fig. 4-2. Most of the masking operations are accomplished within the masking and complement circuit. The *masking circuit* has two inputs: the L or logic register's output which controls the masking function, and the unmasked bus which connects to the gates at the outputs of the various index registers. The output of the masking and complement circuit is the masked bus which connects to the input gates of the several index registers. The *complement circuit* is in series with the masking circuit. It is utilized to change each bit in a data binary word from 0 to 1, or from 1 to 0 whenever the contents of an index register are moved from central control to call store for temporary memory storage. Some of the masking operations are discussed in Chapter 3.

Insertion masking, provided by the insertion masking circuit, is another form of masking that is frequently used. It allows all but a selected group of bits in a specified register to remain intact. The insertion masking circuit is associated with the DBR, as indicated in Fig. 4-2, because insertion masking is often employed when only a portion of a binary word in call store is to be modified. The L register also controls the insertion masking circuit because, in most cases, the bits to be inserted and their associated positions have been recorded in the L register in connection with a previous masking operation.

It should be noted that call store has no circuits for performing insertion or other masking operations. Consequently, masking is entirely a function of central control that is accomplished in two steps. First, the binary word requiring the insertion of information is read from call store.

Next, the required information is inserted in the word and the completed binary word is sent back to call store. For example, assume that an instruction from call store specified that certain data be read out from call store to the X register. The requested binary data word would enter the DBR from call store. It would be gated over the unmasked bus to the masking and complement circuit and then onto the masked bus to the X register. Observe that the data buffer register would contain the data word exactly as received (read out) from call store. The binary word received in the X register, however, may have been modified by the masking and complement circuit if so required by the program instructions. The modified word in the X register, subsequently, may be sent back over the unmasked bus, through the masking and complement circuit, the masked bus, the insertion masking circuit, and then through the DBR to call store. In this case, the modified binary data word could be further altered as it passes through the insertion masking circuit, depending upon the program instructions.

Program Instruction Rudiments

For simplicity, it has been assumed that all instructions originate in program store and that data is stored in call store. During operations, considerable data, especially translation information pertaining to central office codes, are stored in program store. Moreover, the same type of instruction to central control is used to obtain readouts of data from program store and call store. To read (interrogate) either program store or call store for data is a complicated process, particularly since overlap operation of instructions is involved. It is not within the scope of this book to attempt to explain all of the procedures involved. The necessary explanations will thus be presented in a series of simplified steps covering the essential program elements.

Program instructions for electronic switching systems, such as the No. 1 ESS, may be divided into three main categories designated operation, program constant, and index register. In addition, a masking option is included in many instructions. Table 4-1 illustrates this basic composition of a program instructions.

An important operation associated with the program instruction is the indexing process performed by the several index registers. These registers

Table 4-1. Program Instruction Format

Operation	Program Identification Constant	Index Register Address	Masking Option
MX	4	Y	
MY	2	X	PL

are the X, Y, Z, F, J, K, and the data buffer register. Indexing is one method of obtaining an address or location in the call store memory by adding the program constant in a particular program instruction to the binary information previously recorded in one of the index registers.

The indexing process is also used in connection with a list of binary words that may be recorded in consecutive call store addresses or locations. For instance, the program instruction would specify the address of the first binary word in the list. Then, the contents of a designated index register would indicate the position of the desired binary word in this list with respect to the first word. In this manner, the same program instruction may be used a number of times to affect different binary words of a list, thereby reducing the number of required program instructions.

Example of Program Instruction Operations

Central control can go to the index registers for information at any step in a program instruction. For example, the Y register is assumed to contain the address of a block of binary words in call store that are utilized for accumulating the dialed digits of a call in progress. It is required that the reading of the fourth word in the particular block of words be transferred to the X register for further use in the processing sequence. The program instruction for this purpose may be summarized as: Read out the data from the fourth word of a block of binary words in call store whose starting address is now in the Y register, and store this reading (data) in the X register.

The actual instructions to central control from program store are in the form of letter and figure symbols or mnemonics. For the above example, the instructions from program store would be written, ''MX, 4, Y'' as indicated in Table 4-1. The mnemonics, MX, mean memory to X index register. In this respect, the program instructions which specify a read (interrogate data) or write (insert data) requirement in call store memory, contain the letter M in their mnemonic representation, and are designated M instructions. The operating procedure for the aforementioned example, therefore, may be described as follows with reference to Table 4-1 where they are shown under the respective parts of the program instructions:

1. *Operation* = MX (call store memory to X index register). Take data from a specified location in call store memory and store it in the X index register.
2. *Constant* = 4 (binary word with required data). The required data is in the fourth word of a block of binary words in call store whose address is in the index register address data.
3. *Index Register Address* = Y (address information). The starting address, or location in call store of the specified block of binary data words containing the required data, is in the Y register.

Example of Masking Instructions and Operations

To illustrate the application of masking in program instructions, let us assume that central control is directed to ascertain whether or not the first digit dialed was zero (call to the operator). The particular program instructions would be written "MY, 2, X, PL" in order to perform the following tasks:

1. *Operation* = MY. Take data from a specified location in call store memory and store it in the Y index register.
2. *Constant* = 2. The required data is contained in the second word of a block of binary data words in call store at the address in the index register address.
3. *Index Register Address* = X. The starting address or location in call store memory of the specified block of binary data words, containing the word with the desired data, is in the X register.
4. *Masking Option* = PL (Product and logic masking). Mask out all bits in the secondary binary word (constant 2), except the four bits used for the first dialed digit, and store the resultant word in the Y index register. Compare this binary word, using the L (logic) register with the decimal value of 10 (binary 1010). This value is the number of pulses counted when the digit 0 is dialed.
5. *L Register Logic Operations.* If the two compared numerical quantities are equal (first digit dialed is zero), central control will transfer to another type of program at the particular address specified in the instructions. If the two compared quantities are not equal, central control will continue with the existing program sequence because the first digit dialed was other than zero.

The masking and complement circuit and the insertion masking circuit are used to perform the masking options that are specified in the program instructions. In addition to the PL option described in the above example, a PS option is provided by the data-address field of the program instruction. With the PS option, the information in the data-address field is gated directly to the L register, as shown in Fig. 4-3, instead of entering the index adder. The insertion masking and complement options are employed in connection with the transfer of information from a register in central control to an address or location in call store. Two types of insertion masking are normally provided. They are written as EL and ES in the program instructions. Complement masking may be defined as storing a negative number as the complement of the positive number of the same value. It is written as PLC in the program instructions.

Program Instruction Transfers

Addressing or locating instructions in program store is a function of the PAR as previously mentioned. Instructions from program store normally advance in sequence from one particular instruction to the next. This process is controlled in central control by the increment circuit which is associated with the PAR. The increment circuit is comprised of the add-one register (AOR) and the add-one logic (AOL) circuit as shown in Fig. 4-4. Some functions of the increment circuit are illustrated in the following example.

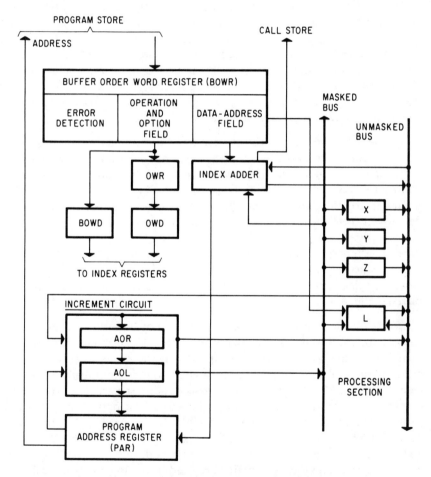

Fig. 4-4. Use of Increment Circuit in Central Control of No. 1 ESS

Let us assume that a certain program instruction, being processed in central control, came from address 206 in program store. In order to receive the next instruction at address 207, it will be necessary to add one (1) to address 206. This add-one function or incrementing operation is performed by the AOR and the AOL component of the increment circuit associated with the program address register (PAR). This incrementing or adding of 1 continues until the completion of the various program instructions involved unless a transfer instruction is received by central control.

The increment circuit also may be utilized with index registers, such as X, Y, and Z, as an option of the program instructions. The incrementing option is often employed when the same processing task is being performed on a number of successive addresses or memory locations. The specific index register is set to the value of the first memory address and the index register is incremented by 1 as successive words are read out from call store. The input and output of the increment circuit, for this function, connect to the unmasked bus and the masked bus, respectively, as shown in Fig. 4-4. Therefore, any index register may be gated to the increment circuit as may be specified in the program instruction.

The increment circuit can be restrained or inhibited by a transfer instruction contained in the program instructions received by central control. A transfer instruction is an instruction from program store directing central control to go to another set of instructions instead of to the immediately following instruction. In such case, the address of the new instruction would be derived from the output of the index adder because it is the place where the contents of the program instructions are combined with the contents of index registers and, consequently, also with memory readings. The index adder connects to the PAR which will generate the new address in program store for this direct transfer of instructions. These interconnections are depicted in Fig. 4-4.

Call store, although normally the source of data for central control, also can be employed as a source of instructions. For instance, a transfer instruction from program store may specify an address in call store for the next instruction. This procedure, designated an indirect transfer, usually is denoted by an M suffix in the index register address part of a program instruction, as shown in Table 4-1. In this event, the particular instruction from call store would be transmitted to the BOWR, per Fig. 4-5, instead of to the data buffer register which interfaces with call store. The BOWR normally interfaces only with program store. In addition, a sequencing circuit is used to assemble the information received in the BOWR from call store because, as later explained, additional time cycles (total of four) are needed by central control to execute program instructions received from call store.

Fig. 4-5. Use of Sequencing Circuit with Decoders in Central Control of No. 1 ESS

Sequencing Circuits

A number of sequencing circuits or sequencers are associated with the decoders, index registers, and other pertinent elements of central control as portrayed by the block diagrams in Fig. 4-5. A sequencing circuit generally performs at least two of the following functions:

1. To inhibit or restrict the operation of decoders.
2. To increase the number of time cycles required for the processing of an instruction from call store or data from program store.
3. To generate the necessary gating signals during the period of the additional time cycles when the decoders are inhibited.

Some of these functions of sequencing circuits are illustrated in the following example.

Assume that an instruction recorded in the OWR specifies that a transfer is to be made to another set of program instructions. It will be necessary, therefore, to immediately restrict the indexing operations in the BOWR because the present instruction in the BOWR is not going to be used. Also, an additional time cycle will be needed for transmitting the instructions contained in the specified transfer address in program store to the BOWR. To accomplish these actions, a sequencing circuit, also termed a sequencer, is used to restrict or inhibit the BOWD and the OWD, and to generate the necessary gating signals during the time cycles when these decoders are inhibited.

A sequencing circuit likewise is required when translation information is read out from program store because this type of information is treated as data and not as program instruction. Thus, a sequencing circuit is needed to provide the three time cycles required for this type of processing. Additional sequencing circuits, as later explained, are provided to perform a number of other functions including the following:

1. Directing program interrupt signals due to trouble conditions or the need to transfer to another processing task, for example, scanning for dial pulses.
2. Repeating the readout of a binary word from program store because of errors that cannot be corrected.
3. Repeating the readout of a binary word from call store because of indicated trouble.
4. Controlling communications with peripheral units.
5. Mutating readouts from a call store that is associated with a signal processor.
6. Stopping and then restarting central control.

Data Instructions from Program Store

In addition to program instructions, considerable translation information or data concerning the routing of calls is stored in program store. Addresses or locations of memory lists (binary words containing instructions or data), as previously mentioned, are part of a joint memory list with certain series of addresses allocated to call store and others to program store. The memory address decoder is utilized to ascertain whether the address recorded in the index adder circuit pertains to call store or to program store. If the address relates to call store, the associated instruction is executed by central control within one time cycle in the normal manner. However, if the address is in program store, the memory address decoder functions to activate a special sequencing circuit. These additional components of central control are shown in Fig. 4-5, which also includes block diagrams of the other components that have been discussed.

The sequencing circuit, in the case where the address is in program store, serves two main purposes. It prevents the translation data, sent by program store from being incorrectly considered as program instruction, and it provides the necessary three time cycles for handling this data within central control. The essential processing steps involved are outlined in Table 4-2 and also explained in the following example.

Let us assume that during the initial time cycle a new program instruction E is contained within the OWR, that its address in program store is 201, and that instruction E requires that certain translation data, including binary word W, be read out from program store. The index adder circuit, therefore, sends address 201 to the PAR for transmission to program store. At the same time, the memory address decoder recognizes the address 201 as being in program store and activates a sequencing circuit to inhibit the BOWD. As a result, the existing instruction A in the BOWR is disregarded. These actions are shown as cycle period 1 in Table 4-2.

In the second time cycle, the requested translation data (binary word W) is received from program store by the BOWR and immediately sent via the index adder and the unmasked bus to the data buffer register which interfaces with call store. At the same time, program store address 321 of program instruction A, which was kept in the add-one register of the increment circuit, is again sent to program store by the PAR. Note, as tabulated in Table 4-2, that program instruction E continues to remain in the OWR during the first, second, and third cycle periods.

During the third cycle period, program instruction E, which had been

Table 4-2. Time Sequence Steps in Processing of Binary Word Translation Data from Program Store

Binary Word from Program Store			Program Store Address
Cycle Period	In BOWR	In OWR	In PAR
1	A*	E†	201
2	W	E	321
3	A + 1‡	E	322
4	A + 2§	A + 1	323

*Initial program instruction at address 321 in program store which was inhibited by action of sequencing circuit because of instruction E.

†Program instruction requesting readout of binary data word W from program store at address 201.

‡Program store instructions at address 321.

§Program store instructions at address 322.

contained within the OWR, is executed just as if the translation data (binary word W which was previously placed in the data buffer register) had been read out from call store instead of from program store. Simultaneously, the next program instruction, A + 1, is received in the BOWR from program address 321 and dispatched to the index adder for necessary indexing operations. At the same instant, the address of the succeeding program instruction, 322 (321 + 1), is placed into the PAR.

Normal program instruction operations are resumed in the fourth cycle period, as indicated in Table 4-2, with the receipt of program instruction A + 2 in the BOWR. In the same period, program instruction A + 1, which is in the order word register (OWR), will be executed and the address of the next program instruction, 323 (322 + 1), is inserted by the increment circuit into the PAR for transmission to program store.

Call Processing Time Cycle and Overlap Operations

The actual circuit actions for each step in the processing of a call take place in central control which, as previously mentioned, is a synchronous machine capable of performing only one action at a time. In the No. 1 ESS processor, these actions are made in a cycle period of 5.5 microseconds. This time cycle is established by the electronic clock depicted in Fig. 4-5. This clock governs the operations of the BOWD and OWD decoders, sequencing circuits, related index registers, and other components. In effect, the clock establishes the gating period for the operation of the various decoders and index registers.

During the 5.5-microsecond period, one of the many thousands of binary words that make up the instructions in program store directs central control in a particular call processing task, or to carry out some automatic maintenance procedure. These different operations follow a precise schedule because central control can execute only one instruction at a time, and each step may require a different time duration. For example, customer lines are scanned periodically to detect the dialing of digits. This scanning must be completed fast enough to ensure that no dial pulses are missed. The switching network, on the other hand, operates at a slower rate so that instructions to make connections in the switching network are issued at a correspondingly lesser interval.

In most cases, a 10-microsecond cycle would be required for each step because of transmission delays in cables, response time of the memory devices in program store and in call store, and other factors as tabulated in Table 4-3. This 10-microsecond period is based on the processing of one instruction at a time by central control (from the preliminary to the execution stages) before another instruction is received from program store. In order to maintain the desired 5.5-microsecond time cycle, central control in the No. 1 ESS is designed to operate in the overlap mode. For this purpose, an instruction is divided into two parts, called the preliminary and execution stages, and central control processes two instructions simultaneously as described below and diagramed in Fig. 4-6.

When a particular program instruction, D, arrives at the BOWR, the PAR sends the address of the following instruction, $D+1$, to program store, and the preliminary stage of incumbent instruction, D, is processed. This overlap operation in central control may be better understood if it is assumed that program instruction D is located at address 251 in program store, and that the format of instruction D is "MX2Z." Therefore, "2Z" is the preliminary stage and "MX" is the execution stage of instruction D.

Table 4-3. Time Period for Call Processing Operations in the No. 1 ESS Central Control without Overlap Mode

Call Processing Operation	Approximate Time in Microseconds
Resetting of instruction register in program store	0.50
Transmissions of instructions from program store to BOWR and the stabilization of its output	2.25
Decoding operations	0.50
Gating of index registers into the index adder and sending the address to call store	1.50
Maximum readout time from call store into data buffer register (B)	3.75
Gating of call store readout from the data buffer register (B) to the specified index register (such as Z)	1.50
Total Time of Operations	10.00

During the preliminary stage of program instruction D, the unmasked bus is used for the indexing operations and any necessary index register modifications are sent over the masked bus to the index adder. These operations are controlled by the BOWR. At the same time, the following tasks are performed within central control:

1. The index adder develops the address, for example, 321, of the required data in call store and sends this address to call store with a readout request.
2. The PAR sends the address 252 $(251+1)$ of the next instruction $(D+1)$ to program store.
3. Program instruction D (comprising format MX2Z in this example) is transferred from the BOWR to the order word (OWR). Therefore, BOWR becomes available to receive the next program instruction, $D+1$, located at address 252 in program store.

In the following execution stage, the data from address 321 in call store is received by the DBR and gated to the Z index register, via the unmasked bus, the masking and complement circuit, and over the masked bus under control of the OWR. At the same instant, the preliminary stage operations of checking, indexing, and index register modifications for the next program instruction, D + 1, that was received from address 252 in program store are performed. Likewise, address 253 of succeeding instruction D + 2 is sent to program store by the PAR. During the overlap time cycle when the masked and unmasked buses are jointly used by the two instructions being processed in central control, the preliminary stage of the next instruction uses only the unmasked bus while the existing instruction in the execution stage uses the masked bus. This arrangement is employed to prevent mutual interference.

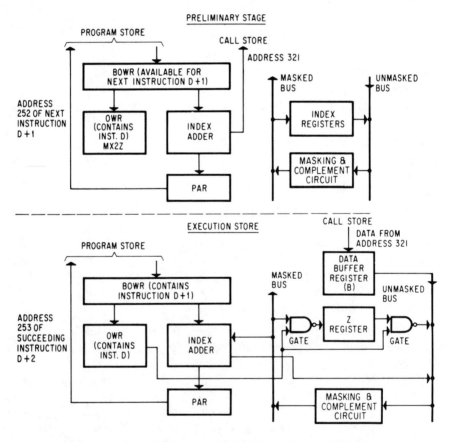

Fig. 4-6. Processing in Preliminary and Execution Stages of Program Instructions in No. 1 ESS

K Register and Accumulator System

To perform the many additions and logical combinations of two quantities that are required in central control, the K accumulator register and associated accumulator circuitry are provided. This accumulator system, illustrated in Fig. 4-7, consists of the K addend and K augend registers, and the logic and arithmetic circuit, in addition to the K register. Also, a rotate and shift circuit and a detect first-one circuit are associated with the K register. The functions of these circuits will be explained later.

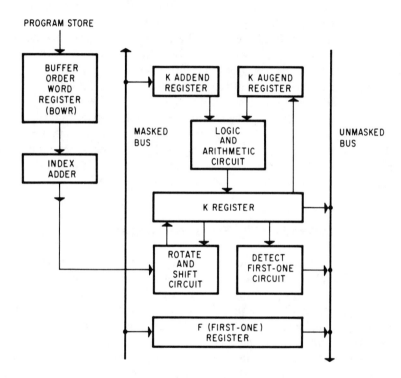

Fig. 4-7. Accumulator System of K Register

The K register, in addition to its use as an index register like the X, Y, and Z registers, also controls various logical and arithmetic operations. For instance, its binary contents can be gated into the K augend register and then combined with the contents of the K addend register in the logic and arithmetic circuit. This combining action normally would employ simple addition or logic functions such as AND, OR, or EXCLUSIVE-OR that were described in Chapter 3.

The binary contents of the K addend register are received, via the masked bus, from the data-address field of the program instruction that had been deposited in the buffer order word register, as depicted in Fig. 4-3. Moreover, when several binary numbers are to be added together, this must be accomplished by adding one at a time. During this process, the partial sums are accumulated in the K register until the final sum is attained.

Binary data in the K register also can be rotated and shifted by the rotate and shift circuit. The shifting action is required to line up two items of information, found in different positions within two binary data words, to a position where the two items may be combined logically. The rotation operation is similar to the shift operation except that, for a left rotate, the contents of the most significant bit are shifted back into the least significant bit instead of being shifted out, and vice versa for a rotation to the right. These rotation and shifting operations are further explained in the section "Logic and Arithmetic Operations in Central Control" in Chapter 3.

The detect first-one circuit is employed to detect and identify a 1 in the midst of a group of zeros in a binary data word in the K register. In effect, it determines the position of the rightmost bit in the K register that is equal to 1. The binary number that identifies the position of this bit is then put into the F or first-one register, via the unmasked bus, the masking and complement circuit, and the masked bus. A circuit to reset this bit in the K register will receive its selection information from the F register.

Functions of CS and CH Flip-Flops and J Register

A pair of flip-flops, designated CS and CH, also connect to the masked bus and to the logic and arithmetic circuit of the K register, as shown in Fig. 4-8. Flip-flops are solid-state devices comprising a combination of logic circuits that provide for the temporary storage of binary digits or bits of information as explained in Chapter 3. The CS and CH flip-flops normally function when a compare operation is being performed in the accumulator system. In this type of operation, the difference of two binary numbers to be compared is computed by the logic and arithmetic circuit.

The result of the computation, however, is not gated into the K register but is utilized only to control the CS and CH flip-flops. These flip-flops also may be controlled during the execution stage of an instruction whenever a binary data word is gated, via the masked bus, to a register other than the K register. In this case, the sign and homogeneity of the binary word will control the CS and CH flip-flops.

These flip-flop and K register functions are related mainly to conditional transfer instructions received from program store. This type of program instruction usually directs central control to proceed to another set of

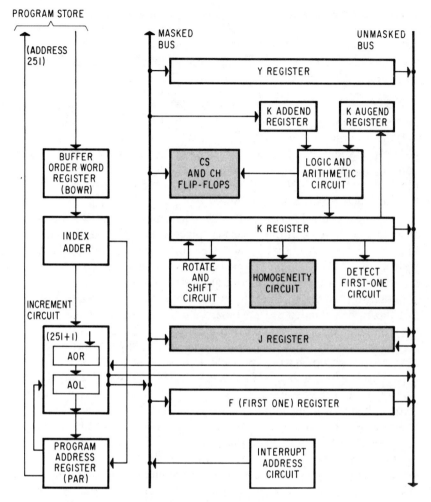

Fig. 4-8. Functions of CS and CH Flip-flops and J Register

instructions, instead of to the immediately following instruction, if a certain data word or bit of data having a predetermined value (such as arithmetic zero) has been deposited in the K register or in the CS and CH flip-flops. If the binary word or bit does not have that value, the transfer is not made and the next following instruction is executed in the normal manner. Arithmetic zero, the usual predetermined value, includes $+0$ bits (all zeros) and -0 bits (all ones). In both cases, all bits are identical and, therefore, the binary word has a homogeneity of 1.

An example of a conditional transfer instruction format utilizing the CS and CH flip-flops is outlined in Table 4-4. The instruction format used

Table 4-4. Format for Conditional Transfer Instruction

Operation	Program Identification Constant	Index Register Address	Masking and Other Options
CW	10	Y	
TCA	251		J

for this example is similar to that shown in Table 4-1. Referring to Table 4-4, the program instruction, "CW, 10, Y" tells central control to compare the program identification constant 10 with the contents of the Y register, and to gate the result to the CS and CH flip-flops. These flip-flops are employed for this comparison operation because the K register is not used for this particular program instruction.

Central control performs the comparison operation by subtracting 10 from the bit contents of the Y register, placing the result on the masked bus, and gating it only to the CS and CH flip-flops. The condition transfer instruction, "TCA, 251, J," which follows is then executed. It directs central control to transfer to the program instruction located at address 251 in program store, if the result of the previous comparison action by the CS and CH flip-flops indicated arithmetic zero. In other words, if the contents of the Y register equaled the program indentification constant 10, arithmetic zero would result. Otherwise, central control would proceed to execute the next instruction which would be at the address in the PAR.

The J option indicated in Table 4-4 is the return address or jump option often associated with transfer instructions. Most transfer instructions frequently occur in the middle of a program. They usually involve a transfer to a subroutine in connection with performing common tasks as explained in Chapter 3. In order to direct the subroutine where to return control to the original program instruction, the J register is provided as shown in Fig. 4-8. When a conditional transfer, as in the foregoing example, is executed, the AOR in the increment circuit contains the contents of the next program store address (251 + 1) to be sent to the PAR. This information, therefore, will be gated to the J register via the unmasked bus. Consequently, when the conditional transfer takes place, the address of the next program instruction will be saved in the J register. Block diagrams of the various components discussed herein are shown in Fig. 4-8 together with the related elements of central control.

Buffer Bus Registers

In addition to the different index registers and the CS and CH flip-flops previously described, a number of other flip-flops are provided in central control of the No. 1 ESS. These flip-flops are arranged in 22 groups, called buffer bus registers, which are associated with the data buffer register. Figure 4-9 shows the arrangement of the buffer bus registers in central control.

The buffer bus registers or flip-flop groups are utilized for maintenance purposes and with match circuits as later explained. Each flip-flop group is assigned a distinct address. Thus, when this address is generated by an instruction from program store, the memory address decoder (illustrated in Fig. 4-5) will operate the gates to and from the particular flip-flop group or buffer bus register. In this way, the control of the buffer bus registers does not require additional space in the program instruction format.

Fig. 4-9. Arrangement of Buffer Bus Registers in Central Control of No. 1 ESS

Interrupt Signals

Interrupt signals may be generated within central control to interrupt a program instruction being executed, and to initiate a transfer to another program instruction associated with the interrupt signal. The interrupt signal is normally originated by the following two conditions:

1. Detection of a trouble condition by the maintenance circuits. This necessitates an immediate interruption in the program being executed in order that the trouble can be analyzed and the required corrective measures be taken before the trouble condition causes errors in call processing operations.

2. Reception of the 5-millisecond clock pulse which initiates the dial pulse scanning program and similar functions that must be performed every 5 to 10 milliseconds.

The interrupt signal causes the interrupt address circuit, which is connected to the masked bus as illustrated in Fig. 4-8, to generate the particular program transfer address required for the type of interrupt signal that was received. A total of ten program addresses are provided for this purpose in the No. 1 ESS. The transfer to the specified address is accomplished under the control of a special sequencing circuit.

The call processing operations in progress within central control are not disrupted when an interrupt signal is received, it is merely postponed. Moreover, the program transfer initiated by the interrupt signal takes place only after the completion of the instruction in progress at the time that the interrupt signal was received. These actions are possible because all pertinent information in the various registers within central control is stored in call store and, therefore, can be retrieved when the interrupted program is resumed. The transfer back to the program instruction that had been interrupted is controlled by a sequencing circuit.

Matching Circuits

In both the No. 1 ESS processor and the 1A processor, two central control units, designated 0 and 1, execute identical tasks within the same program sequences. One central control unit is assigned as the active unit and the other is the standby equipment. The individual data processing steps for each task are closely synchronized because in the No. 1 ESS, the microsecond clock in the duplicate central control unit is driven by only one of the two 2-MHz crystal oscillators. The 2-MHz crystal oscillator in the active central control unit governs the microsecond clocks in both units. In the 1A processor, the clock timing originates from a 20-MHz crystal oscillator circuit in the active central control.

Both central controls start functioning by placing the same address in the PAR to simultaneously obtain and execute the related instructions from program store. Each central control continues the call processing tasks in step with the other. For instance, the same data are read from the call store memory and from the line, junctor, and trunk scanners. Also, identical data processing steps are performed on the received data from call store and each central control unit makes similar decisions. Moreover, certain trouble signals are cross connected between the two units so that if any additional time cycles are inserted in one unit for remedial action, the same number of cycles would be inserted in the other central control unit.

To provide this synchronous type of operation, two match circuits and associated registers are employed within each central control to make continuous checks of the operations between the two units. These checks perform repeated comparisons at certain points in one central control against corresponding points in the other unit for detecting mismatches. For this matching function in the No. 1 ESS processor, two binary data words (one from each central control) are compared in each 5.5-microsecond timing cycle. The binary words used for this matching operation are usually selected from the following processing points within central control depending on the particular program instruction:

data buffer register (DBR)
masked bus
unmasked bus
program address register (PAR)
buffer order word register (BOWR)
index adder
outputs of decoders and sequencing circuits

A simplified block diagram of the matching circuits and related components in the No. 1 ESS is presented in Fig. 4-10. Two internal match buses, labeled 0 and 1, provide access to selected processing points such as the program address register, data buffer register, index adder, and the masked and unmasked buses under control of a decoder in the match circuit. Sampled information from these points is transmitted over the internal match buses to the match registers under control of the gates, and simultaneously to the other central control unit via the external match buses 0 and 1, respectively, as indicated in Fig. 4-10.

Decoders in the match circuits in both central controls normally operate in step with each other. Therefore, the information transmitted from central control 0 to central control 1 is stored in the match registers in synchronization with the action of the gates. The two match circuits in each central control compare the contents of their associated match registers, and generate corresponding output signals in accordance with the presence or absence of a match condition. For example, the detection of a mismatch condition will generate an interrupt signal and further matching operations will be automatically stopped.

The selection of the points to be matched also facilitates the detection of hardware or equipment troubles developing within central control especially during routine tests made in accordance with maintenance programs. Thus, a mismatch condition is detected as soon as the effect of the trouble would be communicated to other elements in the electronic switching system. Signals from the decoders and sequence circuits within each central control also are sent to the respective match circuits as shown in Fig.

4-10. These signals set and reset specific flip-flops which, in turn, govern the selection of internal points within central control for matching during each 5.5-microsecond time cycle.

Fig. 4-10. Central Control Matching Circuits in No. 1 ESS

Similar matching operations are performed between the two central control units in the 1A processor, which is provided in No. 1A ESS and No. 4 ESS offices. Each central control likewise has two separate match circuits. However, the match circuits in the active central control check 24 bits with the 24 internal bits from the other central control during every operation cycle of 700 nanoseconds. These differences in processing time and the number of binary digits (bits), with respect to the No. 1 ESS processor, result from the improved processing structure of the 1A processor. They are further explained in this and subsequent chapters.

Bus Systems and Communications with Peripheral Units in the No. 1 ESS

In carrying out its functions, central control exchanges information (communicates) with its various associated components and peripheral units over common groups of twisted-pair wires called bus systems. Four major bus systems are employed for this purpose. They are designated program store bus, call store bus, central pulse distributor (CPD) bus, and the peripheral unit (PU) bus. Duplicated buses, labeled 0 and 1, link the two central control units with the program and call stores as shown in Fig. 4-11, in order to provide greater reliability.

A simplified block diagram of these major bus systems, which appears in Fig. 4-11, shows the flow of information between central control and the associated elements in the No. 1 ESS. One-half of the program stores and one-half of the call stores are normally installed to the left and right, respectively, of the two central control units in order to reduce the lengths of the respective buses.

The central pulse distributor (CPD) bus and the peripheral unit (PU) bus systems consist of two one-way groups of small gauge twisted-pair wires. The address group of these wire pairs handles the pulses and signals from central control to all units connected to these buses. The answer group of the CPD and PU bus systems conducts the responses from selected peripheral units back to central control. One wire pair is provided in a bus for each binary digit or bit to be transmitted over the bus. A gating method permits a bus to be time-shared by the different units that it serves. The flow of information on any wire pair of a bus usually is in the same direction.

Pulses and other signals usually are applied to a bus in the No. 1 ESS every 5.5 microseconds and they may be only 0.5 microsecond in width. Therefore, in order to minimize propagation delays or transmission time over a bus, it is necessary that the maximum length of wire pairs between central control and the last program store or call store not exceed 125 feet (38.1 meters). For the units on the peripheral unit bus and the central dis-

Fig. 4-11. Central Control Bus Systems in No. 1 ESS

tributor bus, wire pair lengths up to approximately 450 feet (137.2 meters) may be employed without undue propagation delay effects. Note that a total of sixteen central pulse distributors connect to the CPD bus although only a quantity of ten is indicated in Fig. 4-11.

Each wire pair of a bus is connected to a bridged component of the bus through a transformer. Figure 4-12 is a simplified schematic diagram showing this transformer coupling arrangement for a wire pair in the program store bus. The pulse transmitting equipment in central control is connected to the primary of a transformer whose secondary is bridged across the wire pair. In each program store, the pulse receiver connects through a transformer whose primary windings are in series with the wire pair. Each end of a wire pair is terminated in a noninductive resistor, indicated by R-1 and R-2 in Fig. 4-12, in order to maintain the proper line impedance and thereby reduce reflections of current pulses. The center-tapped inductance, L, is bridged across the wire pair to reduce the effects of induced noise by providing a path to ground.

Central control communicates with the sixteen central pulse distributors over the central pulse distributor bus as illustrated in Fig. 4-11. These central pulse distributors are utilized by central control to send orders to or

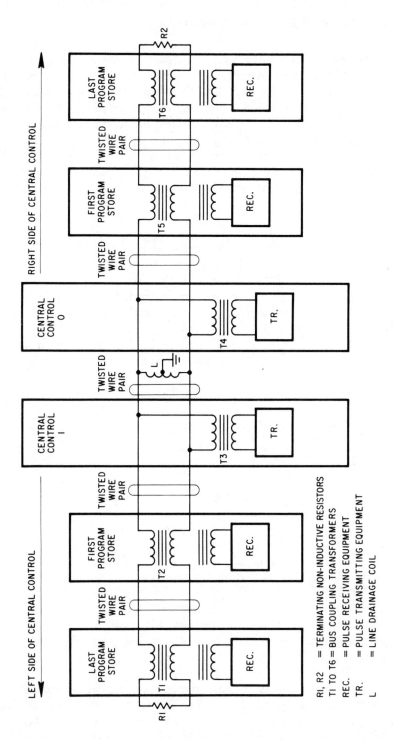

RI, R2 = TERMINATING NON-INDUCTIVE RESISTORS
T1 TO T6 = BUS COUPLING TRANSFORMERS
REC. = PULSE RECEIVING EQUIPMENT
TR. = PULSE TRANSMITTING EQUIPMENT
L = LINE DRAINAGE COIL

Fig. 4-12. Simplified Schematic of Typical Wire Pair Used in Program Store Bus of No. 1 ESS obtain information from peripheral units, such as line scanners, junctor scanners, trunk scanners, network controllers, and signal distributors. For instance, assume that certain line scanning data is needed by central control from central pulse distributor 8. Referring to Fig. 4-13, central control first sends an enable address pulse over the address lead of the CPD bus to all central pulse distributors. Next, central control initiates an execute signal over an individual pair of wires to central pulse distributor 8 which responds by returning a signal over its echo lead. This return or echo signal permits central control to verify that the proper central pulse distributor (CPD 8) has been connected.

Fig. 4-13. Applications of CPD and PU Bus Systems in No. 1 ESS

The selected central pulse distributor (CPD 8) will transmit an enable pulse over a separate direct path to the desired line scanner, 15, as illustrated in Fig. 4-13. This enable pulse causes line scanner 15 to respond to the address signal subsequently transmitted by central control to all peripheral units over the address lead of the peripheral unit bus. Line scanner 15 acknowledges by sending a verify signal back over its same connecting path to central pulse distributor 8. This verify signal is translated by CPD 8 into coded binary digits which are transmitted to central control over the answer lead of the central pulse distributor bus. In this manner, central control can match the returned binary digits from CPD 8 with the original address signal that was previously transmitted in order to determine that only the desired line scanner 15 had been activated.

The information that is sent to a particular central pulse distributor, containing the address of a required peripheral unit, such as line scanner 15, is set up in the F register by appropriate instructions received from program store. The data in the F register is then sent to the central pulse distributor address translator which transmits the proper pulses to the central pulse distributors over the CPD bus as previously explained. The verify answer signal returned from the selected central pulse distributor, for example, CPD 8, is gated into the Y register. This signal in the Y register is then compared in a match circuit with the data in the central pulse distributor address translator. If a mismatch results, an interrupt signal will be generated and a program transfer will be initiated by central control as previously explained.

Information for controlling a particular peripheral unit, such as line scanner 15, is placed in the K addend register and in the K register. The data in the K addend register is decoded into the required binary form by one of the peripheral address translators before being transmitted to line scanner 15. The readout from this line scanner, which is returned over the answer lead of the peripheral unit bus, is gated into the L register for further processing by central control.

Signal Processor Interconnection

In the larger No. 1 ESS electronic switching offices, a signal processor (SP) often is provided to assist central control. A signal processor may be defined as a stored-program processing unit that can efficiently handle input-output (I/O) functions. For instance, the signal processor can detect originating calls, receive dial information from calling customers and process the calls, control the peripheral units of central control, etc. By relieving central control of the aforementioned tasks, the signal processor can increase substantially the traffic capacity of an electronic central office. Two pairs of duplicate signal processors may be provided depending on the size of the central office and its traffic requirements.

The signal processor, in a similar manner as central control, operates under the direction of program instructions. These program instructions, however, are kept in call stores that are associated only with the signal processor. Figure 4-14 shows the relationship of the signal processor and its associated call stores with central control. Up to eight signal processor call stores may be connected with each pair of duplicate signal processors.

Central control and the signal processor in the No. 1 ESS operate

Fig. 4-14. Interconnection of Signal Processor with Central Control in No. 1 ESS

independently of and simultaneously with each other. Central control can always address an SP call store as if it were another call store on its call store bus system. The tasks performed by the signal processor include supervision or periodic scanning of lines to detect originating calls, junctors to detect disconnects, and trunks to check for connections and disconnects. Other actions include the reception of dial pulses or Touch Tone® signals from customers, the reception and transmission of signals with other central offices, and various types of timing operations to detect permanent signals and other malfunctions.

Summary of Central Control Organization in the No. 1 ESS

Many of the principal characteristics and functions of central control in the No. 1 ESS, including its major components, have been described in this chapter. The basic organization of central control also has been explained by a series of step-by-step block diagrams depicting its major components and subsystems. In addition, some aspects of the logic design of central control are described including program instruction operations and transfers, data processing, timing requirements and overlap operations, use of sequencing and matching circuits, and the major bus systems for communicating with associated elements and peripheral units.

Figure 4-15 is a simplified and concise drawing showing, by block diagrams, the major components and associated elements of central control in the No. 1 ESS that are explained in this chapter. Since it is not feasible to delineate in this drawing all of the components and their functions, the

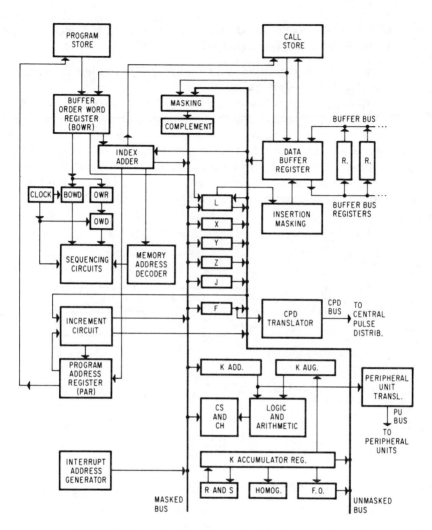

Fig. 4-15. Organization of Central Control in No. 1 ESS

reader should refer to the text and other related drawings for more detailed information.

Central Control Operations In 1A Processor

The general structure of the 1A processor is shown in the block diagram of Fig. 2-17. It is essentially a stored-program control processor that basically consists of a *control system* (explained herein) linked with a *memory system* that is discussed in Chapter 5. The main elements of the control system, which are illustrated in Fig. 4-16, include:

1. Central control, consisting of two fully duplicated units that operate in step with each other and match the results.
2. Input/output unit selectors and the processor peripheral interface (PPI) used for interchanging the binary digit signals from central control with the lower-speed data set signals required by the master control console and local and remote terminals. The PPI also furnishes logic and access circuits for the master control console. Remote terminals connect to distant switching control centers, which monitor and control the operations of several electronic switching offices.

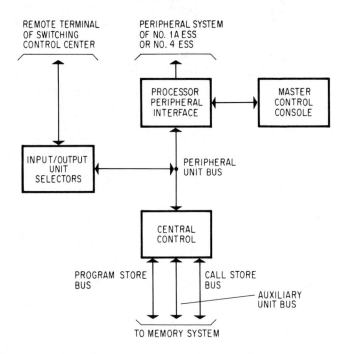

Fig. 4-16. Control System of 1A Processor

3. Communication buses, including the peripheral unit bus, auxiliary unit bus, and the program store and call store buses. These buses convey timing, control, and data binary signals between the various subsystems of the 1A processor. The peripheral unit bus also is compatible with the peripheral bus system in the No. 1 ESS. This makes it feasible to update a No. 1 ESS to the No. 1A ESS type by the addition of the 1A processor and certain related changes.

The memory system, as in the case of the No. 1 ESS, is divided into two main classifications, program store and call store. The instructions in call store have 24-bit words but program store instructions may be either one or two binary words long, that is, either 24 or 48 bits. Program store normally provides the instructions to be executed, and call store furnishes the data. However, either program store or call store can perform these functions because both memory systems communicate with central control in similar manner. The main difference between them is the width of the communication path. Thus, on each readout operation, program store returns a pair of binary words for selection of the required portion by central control, whereas call store sends back only a single binary word of 24 bits. The number of bits in an instruction is not directly related to the execution time. (See Chapter 5 for further explanations.)

The formation of central control in the 1A processor resembles in many ways the organization of central control in the No. 1 ESS as illustrated in Fig. 4-15. Similarly, it executes instructions received from program and call stores, and controls system operations including synchronization and timing. Many improvements have been made to expand and augment the arithmetic and logic functions, instruction features, and register circuits of central control in the 1A processor as well as to greatly increase its processing speed.

For instance, the instruction execution rate of program and call stores, and the various logic circuits, are designed to work with either a 700-nanosecond or 1400-nanosecond operating cycle of central control. Thus, about 750,000 instructions can be executed per second, which is many times faster than the No. 1 ESS central control. Also, *shadow registers* are associated with the general and logic registers in central control (see Fig. 4-18). Shadow registers are used to temporarily store or save data from their related registers for later use in connection with specific program instructions.

Simplified block diagrams of the principal elements of central control in the 1A processor are presented in Figs. 4-17 and 4-18. Some important differences with respect to the structure of central control in the No. 1 ESS (see Fig. 4-15), include the *instruction stack* arrangement for handling instructions received in binary words from progrm and call stores (see Fig. 4-17), and the use of shadow registers illustrated in Fig. 4-18. These registers are associated with *general* and *logic* registers in data-processing operations as subsequently discussed.

Referring to Fig. 4-17, the instruction stack has a design capacity of six binary words, each of 24 bits. A concise explanation of its operation follows. Assume that a double-word binary instruction (48 bits) is read out from program store. It is received by the respective left and right *auxiliary buffer order word registers* (Left ABOWR and Right ABOWR). If both binary words are needed by central control, the contents of the aforemen-

Fig. 4-17. Block Diagram of Instruction Stack in Central Control of 1A Processor

tioned auxiliary buffer order word registers are immediately gated to the corresponding left and right *buffer order word registers* (Left BOWR and Right BOWR), for processing action.

However, should the Left BOWR contain an instruction that has not yet been executed, and if the Right BOWR and the half-word register are available, the contents of the aforementioned Left ABOWR and the Right ABOWR are gated to the Right BOWR and the half-word register (HWR), respectively. But, if only the right half of the binary-word instruction (24 bits) is required, the contents of the Right ABOWR would be gated to the Left BOWR. The different gating paths are indicated by the connecting arrows in Fig. 4-17. This unique instruction stack arrangement permits greater memory capacity in program and call stores, and also substantially reduces instruction fetching and execution times.

The decoding and processing of the binary instructions start as soon as it is available in the Left BOWR. The data portion of a long (48-bit) or double-word instruction would be in the Right BOWR. In the first execu-

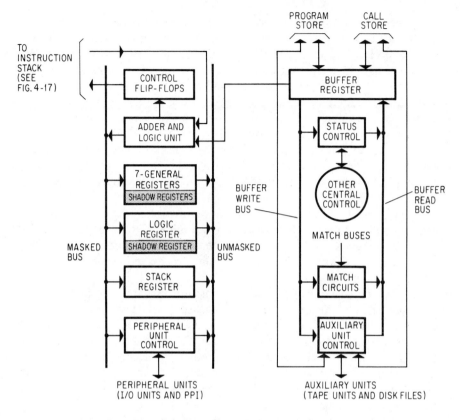

Fig. 4-18. Data Processing and Control Elements of Central Control in 1A Processor

tion cycle, the contents of the Left BOWR are transferred to the order word register, and the data portion of the instruction is gated from the Right BOWR to the address register or to the proper data-processing circuits, such as the adder and logic unit shown in Fig. 4-18. The balance of the instruction decoding is conducted from the order word register (OWR) with the instruction decoder circuits as outlined in Fig. 4-17.

The data-processing and control elements of central control are indicated by the block diagrams in Fig. 4-18. The generic-processing and control operations, such as, masking functions, sequencing, flip-flop circuits, arithmetic, logic, and indexing functions, buffer bus registers, and matching are similar to those described for central control in the No. 1 ESS. The reader, therefore, should refer to Figs. 4-1, 4-2, and 4-4 to 4-11 inclusive, and their accompanying texts for pertinent explanations. They are applicable in many respects to central control in the 1A processor.

Referring to Fig. 4-18, the seven general registers perform arithmetic, logic, and shift roles. The logic register serves both as a masking register and to receive data from peripheral sources. Each of the general and logic registers, as previously stated, has a shadow register to temporarily store data from its associated register for subsequent use. The adder and logic unit (ALU) accomplishes standard logical work besides additions and substractions. The stack register (S) also works with call store in connection with instruction stack operations as illustrated in Fig. 4-17.

A pair of control flip-flops store the sign and homogeneity of a specific binary word after an arithmetic or logical operation. The homogeneity is 1 if all of the bits in a binary word are either 1 or 0; otherwise, it is 0 (see pages 87 and Fig. 4-8).

The *peripheral unit control* circuit monitors and controls the peripheral units (PU). These PU perform the scanning, signal distribution, and switching network functions. Central control communicates with the PU over the peripheral unit bus through the *processor peripheral interface* (PPI) as shown in Fig. 4-16.

The *auxiliary units (AU),* which consist of the tape units and disk files (see Fig. 2-17), connect to central control through the *auxiliary unit control* circuit. However, central control does not have direct access to data stored in these tapes and disks. Requests for transfer of data from tape units and disk files to program or call stores are initiated by the auxiliary unit control as the result of specific program instructions. The above-mentioned auxiliary units then proceed to transfer one binary word at a time through central control under the latter's direction.

Duplicated and matched central controls are provided in the 1A processor as in the No. 1 ESS. The two central controls normally execute identical operations with one central control as the active unit and the other on standby. The status control and match circuits seen in Fig. 4-18, serve to govern these activities. Each central control has two separate match circuits to ensure that both central controls run on equivalent data and execute the same instructions. Every match circuit can compare 24 internal bits to 24 bits from its mate once every 700 nanoseconds to verify that both central controls are running in step. The matching operations are similar to those described in Fig. 4-10 for the No. 1 ESS central control.

The flow of instructions within central control is handled on two internal major bus systems designated *masked bus* and *unmasked bus,* as illustrated in Fig. 4-18. The operation of these bus systems, with their affiliated registers and control circuits, is similar to that previously explained for the masked and unmasked buses in the No. 1 ESS central control.

Questions

1. What three principal parts comprise the central control unit in the No. 1 ESS? What components interface between central control and program store? Between central control and call store?
2. What are the main functions in the No. 1 ESS of: The buffer order word register (BOWR)? The program address register (PAR)? The data buffer register (DBR)? The index adder?
3. Name four or more index registers that are mainly concerned with the processing operations of central control in the No. 1 ESS. What buses connect to these index registers?
4. What circuits within central control provide masking functions? Explain how masking is performed.
5. How many bits are contained in the binary instruction words of program store and call store in the 1A processor?
6. Name three or more functions performed by sequencing circuits.
7. What equipment items comprise the auxiliary units and the peripheral units in the 1A processor? What particular circuits within central control direct their operations?
8. What are the functions of interrupt signals? How are they originated?
9. What functions are performed by the match circuits for the two central control units?
10. What operations are carried out by the general and logic registers in central control of the 1A processor? What type of register is associated with them and what purpose does it fulfill?

5
Memory Systems and Devices

Types of Memory Systems

In processing calls in the No. 1 ESS and the 1A processor, central control is directed by program instructions, translation information, and transient data contained within a memory system as discussed in Chapter 4. In the No. 1 ESS, the memory system (depicted in Fig. 3-1), consists of resident or semipermanent instructions in program store and transient data in call store. The call store memory is utilized to briefly store and then recall or retrieve data pertaining to the processing and routing of a particular called number.

The memory system in the 1A processor, likewise, comprises program and call stores, which perform similar functions as in the No. 1 ESS. In addition, two new memory mechanisms are provided. They are the *file stores* (FS), also called disk files, and the *auxiliary data system* (ADS), which are illustrated in Fig. 5-1. These two memory systems are essential for the direct-memory-access feature of the 1A processor. This function allows interchange of information (binary digits) between the high-speed random access memory (RAM) in program and call stores, and the above-mentioned FS and ADS memory systems.

For instance, various types of maintenance programs can be kept in file stores until needed by program or call stores. Substantial cost savings are realized by this method because of the higher cost of the random access memories in program and call stores. Moreover, the file stores can retain binary data in the magnetic state on its revolving disk without loss during hardware (equipment) troubles and power failures. The transfer of specific programs from file stores to program or call stores for execution, is under the direction of central control.

Fig. 5-1. Organization of Memory Systems of 1A Processor in No. 1A ESS.

File Store and Disk Memory System in 1A Processor

The file stores consist of two disk controllers, each connected over a separate auxiliary unit (AU) bus to central control. A disk controller serves from 1 to 4 rotating magnetic disks as per the block diagrams in Fig. 5-1. The two disk controllers work together but maintain a 180° phase relationship between pairs of disks. Thus, when a request for data is submitted simultaneously to both disk controllers, the first disk to reach the address (location of the binary data on the disk's face) will read out the binary data. The 180° phase difference assures that the read operation will be completed before the other disk can start. Completion of the data readout (job) will cancel this job for the other disk. This 180° relationship also assures that one of the pair of disks always is within a half rotation of any address on the disk's face.

Every disk, which is approximately 24 inches (61 cm) in diameter, is in an airtight enclosure. It is belt-driven by a single-phase 208 V induction motor with a capacitor start winding. This ac motor is powered by a dc/ac inverter that is connected to the -48 V central office battery to assure reliable operations. The frequency of the oscillator, controlling the dc/ac inverter, is changed continuously to control its 180° phase relationship with respect to the other disk in the pair.

There are 100 tracks on each face of the disk so that 200 heads are required, 100 for each face. Each track is divided into 101 sectors, of which 100 are utilized for data and the last sector is used for diagnostic testing. A sector contains 34 binary words (24 bits per word). Consequently, a magnetic disk would have a gross storage capacity of about 16 million bits. A file store can handle 50 separate requests for data during one disk revolution. Therefore, a delay of less than 1 millisecond would result between the data read requests.

Auxiliary Data System in 1A Processor

The auxiliary data system (ADS) also is designed to directly interchange binary data with program and call stores under direction of central control. It stores (writes) data on magnetic tape in contrast with the rotating disks in file stores. In addition, the ADS permits other data input and output devices to be connected to the 1A processor. The main components of the ADS are: data unit selector (DUS), data unit bus (DUB), and the tape frames (TF) as illustrated in Fig. 5-1.

Two pairs of data unit selectors (DUS) are usually installed. The even-numbered DUS (0 and 2) connect to the auxiliary unit (AU) bus 0, and the odd-numbered DUS connects to AU bus 1. Each DUS has access, by means of the data unit bus, to as many as 16 tape frames (TF). Both DUS of a pair access the same tape frame as indicated in the block diagrams of Fig. 5-1.

Each tape frame contains a tape unit controller (TUC) and a tape transport or large diameter reel containing the magnetic tape. The TUC incorporates circuits for directing sequencing operations, error checking, and for maintaining the orderly flow of data to the magnetic tape. These operations are similar to those described for the disk controllers in file stores.

The magnetic tapes in the ADS serve as the system's backup memory. They store system reinitialization and system update programs, and also record call data for subsequent automatic message accounting (AMA) billing purposes. System reinitialization refers to the initial writing or loading of generic programs and office data into program and call stores prior to placing a new electronic office in service. System update concerns the loading of new generic program or office data into the magnetic tape on a tape frame for use as backup memory in the ADS. The memory system continues to function on the original program and office data in the meantime. Upon completion of the system update work, the new program or revised office data is loaded into the appropriate areas of program and call store memories under direction of central control.

The magnetic tape normally used is a 9-track, 800-character-per-inch unit that reads and writes at a speed of 25 inches per second. The fast forward and rewind speeds are 100 inches per second. The equivalent binary data rate (not counting parity checks) is 160,000 bits per second (160 kbs). For dependable operations, the tape unit is powered from a converter that operates from the -48 V office battery.

Memory System Organization

Program instructions, translation information, and transient data stored in the aforementioned memory systems are in the form of binary words. In the No. 1 ESS, binary words in program and call stores comprise 44 and 24 bits, respectively. The program and call stores in the 1A processor are analogous in their hardware structure or framework. The important difference is that program stores respond to read requests from central control with a double binary word (52 bits), while call stores reply with one binary word (26 bits). Note that each binary word, as discussed in Chapter 4, actually consists of 24 bits, plus 2 parity bits for error-checking purposes, or a total of 26 bits. The resident or semipermanent memory in program store is so named because data can be read or retrieved from it as often as necessary by central control. External means, as later described, are employed to write or initially record program instructions and related data that will be retained within program stores. In contrast, call stores have erasable memory storage facilities for temporarily storing data relating to a call in progress. Thus, central control can write binary data into call stores for transitory storage and subsequent readout during the processing of a call.

Central processors in the Bell System's No. 1 ESS, No. 2 ESS, No. 3 ESS, and the 1A processor, which is used in the No. 1A ESS and No. 4 ESS, originally employed magnetic-type memory devices. For example, program stores in the No. 1 ESS utilized permanent magnet twistor (PMT) wires, and ferrod cores were used for the call stores. The 1A processor employed ferrite cores in its program and call store memory systems. Magnetic-type memory systems, likewise, were initially provided by other manufacturers of electronic switching systems making use of the stored-program control

(SPC) concept. The GTE No. 1 EAX, North Electric NX-1E, and Northern Electric SP-1 are some examples in this connection.

During the past several years, many advancements in semiconductor technology occurred, especially in the development of large-scale integrated circuits (LSI) or "chips." These semiconductor devices, by means of the metal-oxide silicon (MOS) fabrication techniques, may contain many hundreds or even thousands of transistors, diodes, and circuit elements, such as resistors and capacitors. Consequently, these chips can perform numerous complex data-processing functions, including serving as flip-flop memory circuits (see Chapter 3). Because of their miniature size, light weight, and low-power requirements, these semiconductor memories are replacing the magnetic core types. This has resulted in substantial cost savings and a considerable reduction of space requirements for electronic central offices.

The aforementioned memory systems are designed to store a binary digit or bit in a specific location in its memory device. And to be able to immediately recall it on demand of central control. It is possible to store several millions of bits in a single memory module (whether of the magnetic core or semiconductor composition) in program or call stores. For instance, bits representing binary instruction words may be kept in program store for years and read many thousands of times daily during call-processing operations. Bits in call store, portraying data for a call in progress, would be temporarily stored and then read during the course of the call. These bits would be erased when new data for a succeeding call is received.

Basic Magnetic Memory Cell

Magnetic memory systems are now being replaced by new and improved semiconductor devices, such as large-scale integrated circuits. It is important, however, that the reader know the principles and methods of operation of the magnetic memory system. It had greatly influenced the design precepts of current electronic switching systems. Moreover, this knowledge should prove very helpful in understanding how the above-mentioned semiconductor chips provide the random access memory (RAM) and related functions in program and call stores. Therefore, the magnetic-type memories furnished in the No. 1 ESS and the improved model used in the 1A processor will be briefly described in this chapter. Succinct explanations also will be presented of semiconductor memory operations.

The basic element of the magnetic memory cell originally employed in both program and call stores of the No. 1 ESS and the 1A processor is a ferrite magnetic core. It is illustrated in Fig. 5-3. This magnetic core consists of a bistable ferrite material of uniform composition, which possesses a square-loop magnetic characteristic shown in Fig. 5-2. It can be magnetized in either a clockwise or counterclockwise direction. But, more important, the magnetization of the ferrite core can be switched very rapidly from one direction to the other.

For instance, assume that the ferrite core is magnetized in a positive or clockwise direction toward point M on the square-loop curve as indicated in Fig. 5-2. It will retain this positive magnetization after the removal of the magnetizing force. To switch or reverse the direction of magnetization, that is, from M to N or in a counterclockwise direction, it would be necessary to apply a negative magnetizing force that can overcome the positive threshold value indicated by R in Fig. 5-2. Similarly, if the core had been magnetized in a negative or counterclockwise direction, it would be necessary to apply a positive magnetic drive to exceed the negative threshold value of S in order to switch the ferrite core's magnetization from the N to the M value shown in Fig. 5-2.

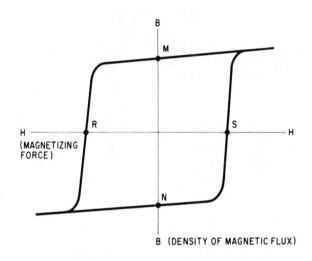

Fig. 5-2. Square-loop Characteristics of a Ferrite Core

Many thousands of these ferrite cores are assembled in a coordinate frame of wires to form a matrix or memory module as used in the No. 1 ESS call store and the original 1A processor. In order to understand the functions of these magnetic ferrite cores in a memory system, let us consider a simple arrangement of a number of these ferrite cores as illustrated in Fig. 5-3. Note that two wires intersect at right angles in the X and Y directions (orthogonally) in the center of each core. The vertical or X column wire is called the X selection wire and the horizontal or Y row wire is called the Y selection wire. Another wire, termed the inhibit wire, passes vertically through the core parallel to the X selection wire. This wire has a continuous

Fig. 5-3. Elements of Basic Memory Cell Utilizing Ferrite Cores

direct current flowing through it as a bias current. This bias current serves to restore the core to its initial magnetic state after the removal of the current pulses from the X and Y selection wires. In call store of the No. 1 ESS, a third vertical conductor is looped through the ferrite cores. It is called the *read* wire, as shown in Fig. 5-14, and is used to conduct the induced voltages as later explained.

Selection of Magnetic Memory Cell

It is essential that direct-current pulses be applied in the required directions through the intersecting X and Y selection wires in order to select a particular memory cell for subsequent recall of its stored binary digit. Consequently, only one-half of the current value needed to change the magnetization of the ferrite core is applied to the X column wire and the other half is sent through the Y row wire. The current in each wire, as a result, will be less than the amount required to change the core's magnetization, for instance, from clockwise to counterclockwise or vice versa. This arrangement prevents the magnetization of a core from being changed if only one of the selection wires passing through it carries current. It also ensures that only the particular ferrite core at the intersection of the current-carrying X column and Y row wires will be magnetized or have its magnetization changed after the current pulses are removed.

Referring to Fig. 5-3, observe that the Y selection wire in the second row has current flowing in the left to right direction. At the same instant, the X selection wire in the third column (going from left to right) has a current pulse in the upward direction. Thus, only the ferrite core located in the third column of the second row will be selected because both X and Y current pulses flow through its core windings. The other ferrite cores in the third column have only X current flowing in the wires through them and only Y current is flowing through the other ferrite cores in the second row. Consequently, there will be insufficient current to magnetize these ferrite cores. The selected core, shown shaded in Fig. 5-3, is magnetized in a clockwise direction because of the direction of the currents flowing through its intersecting X and Y selection wires.

Storage and Read Operations in Memory System of No. 1 ESS

Referring to the No. 1 ESS, assume that the ferrite core of the selected memory cell was magnetized in a clockwise direction, a binary digit 0 would be stored or written into this particular cell. If the core had been magnetized in a counterclockwise direction, a binary digit 1 would be stored instead. This same principle is applicable to the other memory cells in the modules contained in program and call stores. The ferrite core, however, is only one of the elements of the magnetic memory system. Other devices are utilized in both program and call stores as later described.

To recall or read a binary digit that had been stored in the memory cell depicted in Fig. 5-3, the same current pulses are simultaneously applied to the X and Y select wires that intersect in the selected ferrite core. If this core had been previously magnetized in the same (clockwise) direction that would be induced by these current pulses, the core's magnetization will not change. Therefore, no voltage would be induced into the read loop or sensing wire associated with the memory module. This is equivalent to a 0 bit. However, if the applied current pulses are in the opposite direction, the core's magnetization will be switched. This reversal of the core's magnetic flux will cause a voltage pulse to be induced into the read loop for a 1 bit indication. Therefore, to recall or read out binary digits from memory cells may be likened to checking for voltage and no-voltage conditions.

To store or read from a simple memory cell only one bit at a time is not practical for the high-speed requirements of electronic telephone switching systems. Consequently, bits are arranged together in parallel combinations called words so that many bits may be stored in or recalled from the memory module simultaneously. It is possible to form binary words of almost any number of bits by connecting the ferrite cores in certain matrix or wiring patterns. For instance, in the program store memory of the No. 1 ESS central office, each binary word contains 44 bits. In call store, the binary words have 24 bits as later explained.

Organization of Program Store in No. 1 ESS

The memory elements initially used in the No. 1 ESS program store, in addition to the ferrite cores previously described, included a series of twistor wire pairs used for read purposes, copper-strip solenoid loops, and a stack of aluminum memory cards (each approximately 6½ inches by 11½ inches = 16.5 cm × 29.2 cm). These elements were assembled to form the basic memory module of program store. Each aluminum memory card contained 2,880 tiny rectangular or bar magnets made of vicalloy, a special magnetic alloy similar to that used in tape recorders. These minute permanent bar magnets on the aluminum cards were arranged in a grid of forty-five columns and sixty-four rows as illustrated in Fig. 5-4. Every row had forty-five miniature bar magnets of which forty-four were used to represent one binary data word (the last or forty-fifth magnet was not utilized). Each aluminum memory card, therefore, had a capacity of sixty-four binary words (each of forty-four bits) or a quantity of 2,816 binary digits or bits.

A total of 128 of these aluminum memory cards were stacked together with their associated cables of twistor wire pairs, copper-strip solenoid loops, and ferrite access cores to form one memory module which can store up to 8,192 binary words. A package of sixteen memory modules, containing an aggregate of 131,072 binary words (approximately 5,800,000 bits) made up one program store unit. Depending on the number of lines, traffic requirements, and related factors, from two to six program stores usually were installed in a No. 1 ESS central office.

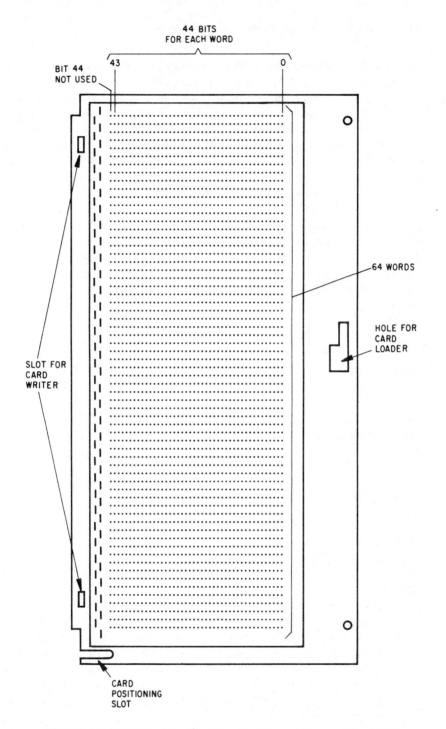

Fig. 5-4. Aluminum Memory Card Used in Program Store of No. 1 ESS

143

The 131,072 binary words in the memory field of a program store were divided into two equal groups or blocks designated H and G. (The "G" and "H" designations were derived from mule driver's terminology, "Gee" and "Haw," for right and left, respectively.) The H half is on the left and the G half is on the right when looking at the front of a program store frame. To ensure the availability of program instructions in the case of power failure, bus or other malfunctions, the H and G halves of a program store were installed on different program store frames. In this way, full duplication of the program memory was provided in the No. 1 ESS.

As shown in Fig. 5-5, the program instructions in the H half of program store 1 are duplicated in the G half of program store 0, and the H half of program store 0 is duplicated in the G half of program store 2. Thus, duplicate program instructions and information always will be available in the adjacent program store. Moreover, the No. 1 ESS can function with an odd number of program stores and also provide full duplication of program instructions.

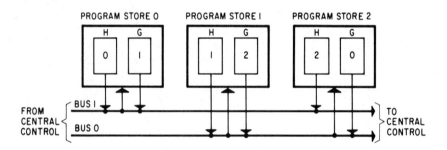

Fig. 5-5. Duplication Method for Program Stores in No. 1 ESS

The H half of a program store memory generally communicates with central control over one bus and the G half uses the other bus. For instance, referring to Fig. 5-5 again, the H half of program store 1 connects to bus 0 while its G half, located in program store 0, utilizes bus 1. Similarly, the H half of program store 2 connects to bus 1 and its G half, installed on program store 1, connects to bus 0.

Consequently, whenever a particular binary word is requested by an address from central control, each of the two program stores containing the word will be requested to read out the word. Both program stores will send back the binary data word to central control, using separate buses. Thus, if one of the program stores should malfunction, the other one with the duplicate information would be able to send it back to central control, utilizing either one or both buses.

Any binary word in program store in the No. 1 ESS can be recalled in less than 5.5 microseconds upon receipt of its address from central control.

In general, approximately 100,000 binary words are employed for the generic or basic program instructions. The number of binary words needed for translation information depends on the number of trunks and customer lines. Approximately three binary words are required as translation information for each line. Means also are provided to ensure that stored data within program store will be protected from accidental destruction by equipment malfunctions or operator errors.

Twistor Wire Pairs and Copper-Strip Solenoid Loops in Program Store

The recall of binary data stored in the program store modules of No. 1 ESS is accomplished by a *twistor wire pair or loop*. A twistor wire is a length of very small gauge copper wire which is helically wrapped with a thin permalloy tape. A plain copper wire of the same gauge is run parallel to the twistor wire and connected to it at one end to form a read pair or loop. The other end of the pair connects to a sensing amplifier and other circuits in program store required for read purposes. Figure 5-6 illustrates the basic twistor wire loop used in the memory module. A total of forty-five twistor wire pairs are assembled into a flat plastic cable. Two such cables are utilized for each memory module, one on each side of the solenoid loops as described below.

Fig. 5-6. Twistor Wire Pair Used to Read Out One Binary Digit in No. 1 ESS

The main part of the memory module consists of a vertical stack of sixty-four *copper-strip solenoid loops,* each of which passes through the center of a small ferrite core attached to an insulating board. The two aluminum memory cards face this stack of sixty-four copper-strip solenoid loops, one card on each side of the stack. Two forty-five-pair twistor wire cables are installed perpendicular to the respective sides of the solenoid loops so that each cable is between the stack of sixty-four solenoid loops and an aluminum card. Figure 5-7 is a simplified sketch of this assemblage.

A twistor wire pair, therefore, will traverse the same binary digit or bit position in every one of the sixty-four rows of binary words represented on an aluminum card; that is, one of the miniature bar magnets on each alumi-

Fig. 5-7. Memory Module Section in Program Store of No. 1 ESS
(Courtesy of Bell Telephone Laboratories)

num card will appear at every intersection of a twistor wire pair and a copper-strip solenoid loop. Since one aluminum memory card appears on each side of a stack of solenoid loops, the sixty-four insulating boards, with their stacks of solenoid loops that comprise a memory module, are used to retrieve binary words on the 128 aluminum cards.

Selection of Ferrite Access Core in Program Store

A complete memory module in a program store of the No. 1 ESS as previously described, consists of sixty-four separate sections of solenoid stacks. Every section contains a stack of sixty-four solenoids and ferrite access cores. Consequently, there will be a total of 4,096 ferrite cores in a memory module arranged in a sixty-four by sixty-four matrix or array. The sixteen memory modules in one program store make up 65,536 ferrite access cores.

The two common vertical wires, the bias and X selection leads, and the Y selection lead wire run through the core at right angles to each other. The X and Y wires actually form two turns around the ferrite core. The single bias wire has a direct current continously flowing through it.

To initiate a recall of a specific binary word from the sixty-four words stored on each of the 128 aluminum memory cards positioned in a memory module, it first is necessary to select the proper ferrite access core. This selection is governed by the address information received from central control in the form of binary digits, via the program store's control circuits.

For the selection process, the H and G halves of each program store are assigned a four-digit binary name. This name contains four binary digits, two binary ones and two binary zeros, for example, 1001. In addition, a 16-bit address code is used to identify the desired binary word from the 65,536 binary words stored in each H and G half of a program store. These binary digits indicate the location of the desired word by its particular memory module and H or G designation. This address data also is sent from central control.

The main elements of a program store's control and core access selection circuits are shown in the block diagram of Fig. 5-8. The four binary digits of the name designation are compared by the decoder unit with the related binary digits assigned to the H and G halves of program store. If there is a match, the decoder will generate a binary 0 digit for the H half or a binary 1 for the G half. The particular binary digit or bit is immediately sent to the address register. Therefore, seventeen bits will be contained within the address register.

The address data recorded in the address register will cause the core access circuits to connect to the proper X and Y access switches. These switches, in turn, will select the required X and Y diode matrices in order to determine one of the 256 X selection wires and one of the 256 Y selection wires which will intersect in the desired ferrite access core. This process is illustrated in Fig. 5-8.

Binary Digit Read Process in No. 1 ESS

When the desired ferrite access core is selected, the control circuits in program store initiate a 2-microsecond dc pulse simultaneously over its intersecting X and Y selection wires. The combined X and Y pulses produce

Fig. 5-8. Selection of Ferrite Access Core in Memory Module of Program Store in No. 1 ESS

a magnetic field exceeding that generated by the continuous direct current flowing in the bias wire. Thus, the magnetic state of the selected ferrite access core will be changed. A current pulse, therefore, will be induced in the copper-strip solenoid loop associated with the selected ferrite core.

The induced current pulse in the solenoid loop actually causes the interrogation of two 44-binary digit words at the same time. One binary word is stored on the adjacent row of forty-four bar magnets located in the aluminum card on the left side of the solenoid loop; the other binary word is on the same row on the aluminum card on the right side. However, only one binary word will be sent to central control in accordance with the address information received by program store.

The binary read process involved may be better understood if we examine one of the bar magnets in the row, and its associated twistor wire pair at the intersection of the specific solenoid loop, as shown in Fig. 5-9. In this connection, recall that each bar magnet is either magnetized or demagnetized. A magnetized vicalloy bar magnet represents a stored binary 0 digit. A demagnetized vicalloy bar corresponds to a binary 1 digit.

Fig. 5-9. Intersection of Bar Magnet, Solenoid Loop, and Twistor Wire Readout Pair in Program Store of No. 1 ESS

Referring to Fig. 5-9, let us assume that the first vicalloy magnet in the row is not magnetized. The interrogating current pulse in the solenoid loop, therefore, will cause a magnetic polarity change in that part of the twistor wire's permalloy tape, which is located at its intersection on the solenoid loop with the unmagnetized bar magnet. This change in the magnetic flux will induce a voltage pulse in the twistor wire. No voltage, however, will be induced in the plain copper wire of this twistor wire pair. This plain wire provides a return path for the read current through the external connected circuits. Consequently, the induced voltage in the twistor wire pair will result in a binary 1 readout for the unmagnetized bar magnet in this row on the aluminum memory card. Moreover, any induced noise voltage will be in the same direction in both the twistor and plain wires of the readout pair and, therefore, will cancel out.

When the current pulse is removed from the X and Y leads, the specific ferrite access core at their intersection will be restored to its initial magnetic state by the continuous direct current in the bias wire. This action will produce a current pulse in the respective solenoid loop which will be opposite in direction to the initial interrogate pulse. Thus, the portion of the permalloy tape at the intersection of the twistor wire and the solenoid loop will change back to its original magnetic polarity.

Let us now assume that the next vicalloy bar magnet in the same row on the aluminum card is permanently magnetized. In this case, the magnetic field caused by the interrogating current pulse in the solenoid loop will add to the magnetic field of this magnetized bar magnet. Therefore, there will be no change in the magnetization of the twistor wire opposite this bar magnet. As a result, little if any voltage will be induced in the respective twistor wire pair so that the readout will be a binary 0.

At the completion of the current pulses on the X and Y leads, the ferrite access core will also revert to its original magnetic state. The resultant induced voltage pulse in the solenoid loop now will generate a magnetic field which is opposite to, but weaker than, the magnetic field due to the permanent vicalloy bar magnet. As a result, the permalloy tape of the twistor wire pair will retain its initial magnetic polarity.

Recall of Binary Word from Program Store

The foregoing read process is applied simultaneously to the forty-four operative vicalloy bar magnets on the same row of both aluminum cards facing the chosen solenoid loop. Therefore, either a binary 0 or a binary 1 signal will ensue in each of the forty-four associated twistor wire pairs. The simultaneous readout of forty-four binary digits comprises one binary word.

The method of handling the binary word readouts can be seen by referring back to Fig. 5-7. The forty-four twistor wire pairs in Cable A will convey the 44-bit binary word readout from the left aluminum memory card. In a similar manner and at the same time, the different binary word readout from the aluminum card on the right side of the solenoid stack will be carried by the forty-four twistor wire pairs of cable B. Selection circuits in program store, as indicated in Fig. 5-10, are utilized to select and amplify the desired binary word in accordance with the address information received from central control.

To reduce circuit complexities and the number of individual amplifiers required, the read leads of the sixteen memory modules in a program store are paralleled in groups of four. Each of the four modules in a group is picked from a different location in program store to minimize electrical noise interference. Figure 5-10 is a block diagram indicating a typical division of the sixteen memory modules into four groups, designated 0 to 3, inclusive.

The read pairs of the four similar numbered modules in a group are connected in parallel as shown for the modules designated 0 in Fig. 5-10. Groups numbered 1, 2, and 3 are connected in a similar fashion. Thus, four groups of eighty-eight readout pairs each are connected to the selection circuits where the desired group of forty-four read pairs is chosen.

A group of forty-four read amplifiers provides the necessary gain to the outputs of the selection circuits (as shown in Fig. 5-10) in order to properly operate the sampler unit. The sampler unit makes the decision

Fig. 5-10. Read Selection and Connection Circuits for Memory Modules in Program Store of No. 1 ESS

whether the output is binary 1 or binary 0. It also quantizes or prepares the output pulse to be a length of one-half microsecond for subsequent transmission over the connecting bus to central control. The output of the sampler, therefore, will contain a series of forty-four binary one and zero signals which form the desired binary word of the program instruction for use by central control.

Changing Information in Memory Cards in No. 1 ESS

There are, as previously explained, 2,048 aluminum or twistor memory cards in a program store, divided into 16 modules of 128 cards each. These memory cards contain the program instructions and translation information for directing the operation of central control. It is not often necessary to change the program instructions after an electronic central office is placed in operation. Translaion information, such as trunk routings and status of customer lines, however, requires frequent modifications. Some of this information, relating to recent changes in customer lines and trunks, is temporarily stored in call store prior to updating the data in program store. Central control will refer to the translation information in program store if the desired information is not found in call store. It is necessary, usually at monthly intervals, to prepare a new set of twistor memory cards which will incorporate changes that have been made in the translation information.

The equipment originally used in No. 1 ESS offices to effect translation changes in program store consists of a memory card writer and two card loaders together with a set of 128 spare cards (see Fig. 5-11). The 128 twistor memory cards in a module are handled as a unit for effecting the translation changes. The cards are withdrawn from, or inserted into, a memory module by one of the motor-driven card loaders which also is used to transport them to the card writer unit. Recall that the sixty-four right-hand and the sixty-four left-hand twistor cards in a module have their mag-

Fig. 5-11. Memory Card Writing Equipment Used with Program Store in No. 1 ESS

nets facing each other. Consequently, the magnets in one-half of the cards will be face up and the magnets in the other half will be face down. In order that all cards will be positioned so that their magnets will face up during the information writing process, two mounting positions are provided on the card writer unit, designated Pass A and Pass B.

To update translation information in a program store, a memory module is first selected in accordance with information received from central control over the teletypewriter at the master control center. The card loader, containing the spare set of 128 twistor memory cards, is mounted on the Pass A position of the memory card writer and the appropriate keys are operated to initiate the writing operations. The memory card writer unit obtains the new translation information from central control, one binary word at a time as needed for the writing process, and stores it in a register circuit. The card writer mechanism next withdraws one memory card at a time from the card loader for the writing process.

The new translation information is written on the memory card, one 44-bit word at a time, by passing a special type of writing head across the surface of the card. A 20-kHz sine wave current is sent through a winding on the writing head. The miniature bar magnets in the memory card are magnetized or demagnetized by altering the waveform of the 20-kHz current in accordance with the binary data that was stored in the register circuit of the memory card writer. Upon completion of the writing process, which takes a matter of seconds, the memory card is returned to the card loader.

The writing process continues until all sixty-four right-hand cards have been completed. Then the card loader is manually inverted to the Pass B position and the remaining sixty-four left-hand cards are processed in the same manner. Upon completion of the card writing work on the 128 cards, a program store in which a memory module is to be updated is removed from service. The memory cards are removed from this module by the second card loader and the newly written cards in the first card loader are immediately inserted in the module. The translation information on the new cards is then verified with central control and, if found to be correct, the program store is returned to normal service. This sequence of steps is repeated to rewrite the cards in the memory module containing the duplicate of the particular translation information. The entire process is repeated as required for the other program stores.

Call Store Memory Systems

The call stores furnish the temporary read-write or erasable memory systems in the No. 1 ESS and in the later electronic offices, such as the No. 1A ESS and No. 4 ESS, which employ the 1A processor. Unlike program store, information may be temporarily recorded in call store and then read out, altered, or erased as directed by central control. The information stored in the temporary memory of call store relates primarily to the handling of calls in progress within the electronic central office.

The type of information includes, for example, the network terminations used for each call in progress, the dialed digits received for the called number, the specific digits to be outpulsed or sent to the distant called office, the busy or idle status of customer lines, trunks, junctors, and network links, customer billing information to be recorded on the automatic message accounting (AMA) tape, and recent changes in data concerning customer lines and trunks prior to updating the associated translation information in program store.

Duplicate input and output bus systems are provided to ensure high-speed reliable communications between call store and central control. These bus systems also control high-speed switching of duplicated memory blocks within call stores. Each call store communicates over either or both duplicated buses as directed by central control. All call stores share the same set of signaling and addressing leads. Furthermore, special maintenance control modes are provided for the interrogation of various test points within call stores, and to permit the alteration of operational features of internal circuits as may be required by central control.

Organization of Call Store in No. 1 ESS

Information stored in a call store of a No. 1 ESS is contained in binary data words of twenty-four bits each. One of these bits, however, is utilized for parity checking purposes. Each binary word has a specific location in call store which is identified by an address sent from central control. The particular mode to be performed, for instance, write or read, is specified by central control. For a write mode, central control also sends the binary word to be written in call store. In case of a readout request, the required binary word is located and read out for transmission to central control. The interval between consecutive operations in call store of the No. 1 ESS is 5.5 microseconds.

There are 8,192 binary word locations in one call store of the No. 1 ESS. This provides a storage capacity of 196,608 bits on the basis of twenty-four bits per binary word. The number of call stores installed in an electronic central office depends on the number of customer lines and the busy-hour calling rate. For example, a 10,000-line central office with a busy-hour calling rate of 13,000 calls would usually require four call stores including the two duplicate ones. Large central offices may require as many as thirty-six call stores. If signal processors are installed, each pair of signal processors can have up to eight call stores.

Call store functions may be divided into four main categories: memory module, associated memory circuits, logic and sequence circuits, and maintenance circuits. Figure 5-12 diagrams these main functions.

There are four memory modules in a call store of No. 1 ESS. Each module has 2,048 binary word locations or a total of 8,192 binary words in the call store. The memory modules contain the ferrite sheet cores as later explained. They are divided into two information blocks designated the H half and the G half. This division is similar to that created in program store.

Referring to Fig. 5-12, the input bus selection circuit connects call store to one of the duplicated buses from central control. The output bus selection circuit similarly selects one of the output or answer buses used to send the read binary word to central control. The memory circuits, which surround the four memory modules, control the access, inhibit, and read operations. The access circuits are directed to the desired binary word loca-

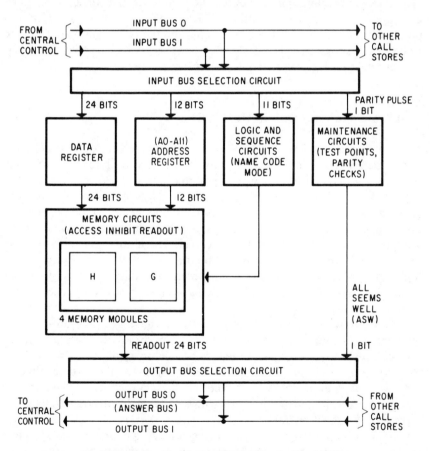

Fig 5-12. Major Elements of Call Store in No. 1 ESS

tion in a particular memory module by a 12-bit address word from the address register. The selection of the specific H or G half is determined by the binary digit derived from the 6-bit name code that is sent to the logic and sequence circuits. The data register holds the 24-bit data pulses, which were received from central control, for subsequently writing a binary word into call store. This register also is used in the read mode to hold the particular recalled binary word until it is written into the memory module.

The logic and sequence circuits decode the 2-bit read or write order, the 6-bit name code, and the 3-bit mode order words. One bit, as previously mentioned, is derived from the name code and used to select the H or G half of the memory module. The 3 bits in the mode order determine whether or not the particular call store should respond and the mode of operation, that is, whether a binary word is to be read out from or written into the memory module.

The maintenance circuits check the validity of the single parity pulse generated by the name code and the address. These circuits also verify test points within call store. If these test points and the parity pulse are found to be correct, a single bit pulse, designated "All Seems Well" (ASW), is generated and sent back to central control over the output bus.

Call Store Memory Cell in No. 1 ESS

The ferrite sheet with its 256 holes or cores is the basic storage element in the memory modules of a call store in No. 1 ESS. Each ferrite sheet is approximately one-inch (2.54 cm²) square and about 0.030 of an inch (0.08 cm) thick. The sheets are arranged in three adjacent stacks of sixty-four sheets each, or a total of 192 ferrite sheets in one memory module. A single ferrite sheet contains 256 holes, each 0.025 (0.06 cm) of an inch in diameter and placed on 0.050-inch (0.13 cm) centers. These holes are formed into a 16 by 16 grid or matrix with each hole providing storage for one binary digit. They may be considered as equivalent to an array of 256 miniature magnetic cores. Figure 5-13 illustrates a typical ferrite sheet used in the memory module.

The equivalent cores in the ferrite sheet have square-loop magnetic characteristics similar to the ferrite cores in program store which were explained previously and shown in Fig. 5-2. One binary digit can be stored in the ferrite material surrounding each hole by magnetizing it in either a clockwise or counterclockwise direction. A clockwise magnetization will store a binary 0, and a binary 1 will be stored by a counterclockwise direction of magnetization.

The holes in the ferrite sheet, as previously explained, correspond to a ferrite core memory cell as used in program store. Four conductors are as-

Fig. 5-13. Ferrite Sheet Used in Memory Module

sociated with each hole in the ferrite sheet which may be considered as a memory cell (see Fig. 5-14). They are the X and Y selection wires, the inhibit wire, and a read wire. The Y select wire actually is a printed conductor on both sides of the ferrite sheet. The printed wiring also forms a metallic sleeve through the hole to connect together the printed conductors on the opposite sides of the ferrite sheet. Each row of holes is connected to the next row so that all holes in a ferrite sheet connect to the same Y select wire. The X and Y selection wires carry the dc pulses to access the desired hole or memory cell in a similar manner as described for the ferrite core in program store. Pulses representing binary digits read from the memory cell are induced in the read or sense wire. The inhibit wire is utilized to prevent a

READ WIRE
INHIBIT WIRE
X SELECT WIRE
Y SELECT WIRE ON
FERRITE SHEET

Fig. 5-14. Equivalent Ferrite Memory Core Used in Call Store of No. 1 ESS memory cell from switching or changing its magnetization polarity during the write operation. For this purpose, a dc pulse is sent over the inhibit wire simultaneously with the current pulses over the X and Y select wires, but in the opposite direction.

Each hole in the ferrite sheet, as previously mentioned, can store one bit in its surrounding material. To handle the 24 bits comprising a binary word in each call store's module, the sixteen rows of holes in the ferrite sheet are subdivided into eight row pairs. This results in 8 bit levels or planes. The read or sensing wire is coupled through two adjacent rows for each bit. Thus, by utilizing three stacks of bit pairs, the required 24 bits in the binary word will be collated.

Read Operations in Call Store of No. 1 ESS

Read and write operations in call store of the No. 1 ESS are carried out in two stages. The first stage in the read operation is to gain access to the location of the required binary word in the memory module. This is accomplished by sending a current pulse simultaneously over one of the sixty-four X selection wires and one of the sixty-four Y selection wires in the specified H or G half of the particular call store. These current pulses are bipolar, that is, a negative current pulse is first sent and then immediately followed by a positive pulse. The negative pulse is used to read the memory cell which is at the intersection of the specified X and Y selection wires.

The combined current pulses over the X and Y leads will reset the particular memory cell, that is, its magnetic state will be switched to that which is associated with binary 0. If the memory cell already is in the binary 0 state, there will be a negligible change in magnetization. Consequently, a very low output voltage will be induced in the read or sensing wire. Now, if the memory cell's initial magnetic condition should correspond to the binary 1 state, the aforementioned combined X and Y current pulses will cause a change in its magnetization. This magnetic change will induce an appreciable voltage in the read wire. Thus, no voltage in the read wire indicates that the memory cell initially was in the binary 0 state, while the presence of a voltage indicates an initial binary 1 state. The resultant binary digit outputs are stored in the data register illustrated in Fig. 5-12.

Irrespective of its initial state, the memory cell is always changed into the binary 0 state by the read operation. Consequently, the memory cell in call store is considered to have a destructive read operation. It is essential that the information previously stored in the memory cell not be lost. Therefore, during the second stage of the read operation, the circuits in the memory module are designed to write back whatever was read out.

For example, to write back a digit, the location of the previously recalled binary word again is selected by pulsing the same X and Y selection leads. But the direction of the current pulses now is opposite to that used for the read operation. Therefore, each selected memory cell will be set, that is, switched from the binary 0 state to the binary 1 state. Consequently, the X and Y current pulses would tend to write back a binary 1 in each bit of the selected binary word location. However, any bit in the data register that is equal to 0 will cause an associated inhibit circuit to function. This circuit will apply a negative current pulse to the related inhibit wire. The resultant magnetic field, due to the inhibit wire and the opposing magnetic flux of the X and Y selection wires, will not be sufficient to switch the memory cell. Accordingly, the memory cell's magnetization is not changed and it will remain in the binary 0 state.

Write Operations in Call Store of No. 1 ESS

To write new information in the form of binary words into a memory module of call store, a read operation is first performed on the memory cells of the binary word in the specified location of the H or G half of call store. This is necessary so that all 24 bits of the selected binary word will be switched to the binary 0 state if any should be in the binary 1 state. The bits which are read out, however, are not stored in the data register as in the case of a normal read operation.

The effect of the first stage of a write operation, therefore, is to clear or reset to zero the binary word in the specified location. Next, the information to be written is received over the input bus from central control, in the form of a 24-bit binary word, for storage in the data register. During the second stage, this information is taken from the data register. It is written into the selected binary word location on the ferrite cores of a particular memory module in the manner previously described.

For example, to write a binary 1, positive pulses are applied to the appropriate X and Y selection wires to create a counterclockwise magnetic field in the selected memory cell. Memory cell A in Fig. 5-15 illustrates this condition. The effect of this action is to reverse the magnetization of the cell which was left in the binary 0 state by the read operations during the first stage. As a result, a binary 1 will be stored in this particular cell.

A binary 0 is written into a memory cell by not changing its magnetic state from that resulting from the read operations during the first stage. To achieve this, a negative current pulse is applied to the inhibit wire at the same time that the positive current pulses are sent through the X and Y selection wires. Memory cell B in Fig. 5-15 indicates this condition. The resultant magnetic field will not be sufficient to switch this memory cell's magnetization from a clockwise direction and it will remain in the binary 0 state. Memory cells that are not selected by coincident current pulses in the X and Y conductors retain a weak remanent magnetic field. Memory cell C

Fig. 5-15. Writing Binary Digits in Memory Cells of Call Store of No. 1 ESS

in Fig. 5-15 illustrates this situation for a current pulse only on the Y conductor of an unselected memory cell. Memory cell D represents a current pulse in the upward direction on the X wire only of a similar unselected cell.

Memory Cell Access Circuits in Call Store of No. 1 ESS

A particular memory cell in a ferrite core in a module of call store is selected by its associated X and Y access circuits. Figure 5-16 is a schematic drawing of the selection arrangements. The access circuits are controlled by the 12-bit address information received from central control and recorded in the address register as shown in Fig. 5-12. The first six of the 12 address bits (A0 to A5) control the Y access circuits. The other six address bits (A 6 to A 11) similarly control the X access circuits.

Two X and two Y access circuits are provided to select the desired memory cell in a memory module. One X and one Y access circuit are associated with the H half of the module and, likewise, one X and one Y access circuit are utilized with the G half. Thus, an access circuit can select one out of sixty-four memory cells in the ferrite cores in the desired H or G half of the memory module.

The X and Y access circuits are connected through a number of electronic switches to dc pulse generators or drivers as shown in Fig. 5-16. Separate read and write current generators are provided to pulse the X and Y selection leads, first in one direction and then in the opposite. Each electronic switch consists of solid-state devices, such as integrated circuits, which allow current to flow only when a particular combination of address bits is present. Each electronic switch requires a different combination of three address bits for operation.

For instance, referring to Figs. 5-12 and 5-16, during a read operation, current can flow through only one "A" switch at a time to the associated X access circuit whenever the specified combination of three address bits (in the A6 to A11 sequence from the address register) is received. The desired "B" switch is controlled in a similar fashion to connect to the X access circuits in connection with a write operation. The Y access circuits are associated with the "C" and "D" electronic switches for read and write operations, respectively. A combination of three address bits in the A0 to A5 sequence received from the address register, is utilized to control these switches which function in the same manner as described for the "A" and "B" electronic switches. In this way, a current pulse will flow through only one X wire out of sixty-four, and through only one Y conductor out of sixty-four in a particular memory module of call store. The same X and Y wires will be pulsed in the opposite direction during a write operation.

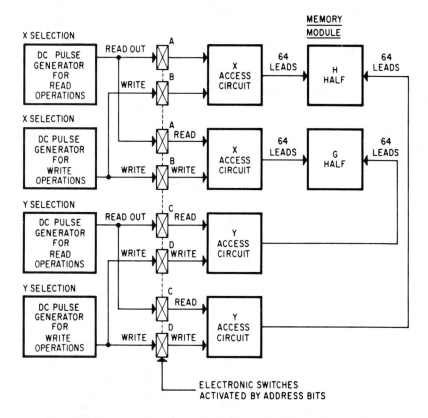

Fig. 5-16. Memory Cell Access Methods in Call Store of No. 1 ESS

Duplicate Input and Output Operations of Call Stores in No. 1 ESS

Requests from central control for read or write operations are accomplished simultaneously by two different call stores. The H and G halves of the information blocks in every call store are assigned a 6-bit name code as previously explained. The H half's code name is established by appropriate wiring at the time of installation. On the other hand, the G half is assigned a code name by the action of its flip-flop circuits under direction of central control. This coding procedure permits selected call stores to be connected to central control utilizing a common bus system.

Figure 5-17A illustrates a typical connection of four call stores, containing information blocks 0, 1, 2, and 3 to the duplicate input and output buses. For instance, assume that central control is requested by a program to read binary word A which is located in information block 2. Two copies of this binary word are available, one in the G half of call store 1 and the other in the H half of call store 2. Central control identifies the desired binary word by means of an 18-bit address code. This address code is divided into a 6-bit name code, K, which indicates the location as in information block 2, and a 12-bit address, A, which specifies the particular location of binary word A within block 2.

The name code, K, and the address, A, are transmitted by central control over both read buses 0 and 1 to all call stores. However, only call stores 1 and 2 will detect a match between the K name code and their internally assigned name codes. Consequently, only these two call stores will utilize address A to read out the desired binary word. For example, the G half access circuit would be activated in call store 1 because its name code matches the K name code that is received. Similarly, the H half access circuit in call store 2 will be activated. Moreover, bus 1 transmits the address data and receives the bits read from binary word A in the G half of call store 1. Bus 0 supplies the address data and receives the bits read from the duplicate binary word A located in the H half of call store 2. The same procedure is followed for all other information blocks in call stores.

The duplicate buses and the control circuits within call stores provide means for overcoming trouble conditions that may be encountered in call stores or its interconnecting buses with central control. For instance, call store 1 has been taken out-of-service because of a trouble condition. The flip-flop circuits of call store 1, therefore, will reject any requests for its information from blocks 1 or 2. Consequently, only the G half of call store 0 will be available to handle requests from central control for information from block 1. The readouts from this G half, therefore, will be sent on both buses 0 and 1 as shown in Fig. 5-17B. For the same reason, only the H half of call store 2 will be able to handle requests for information from block 2.

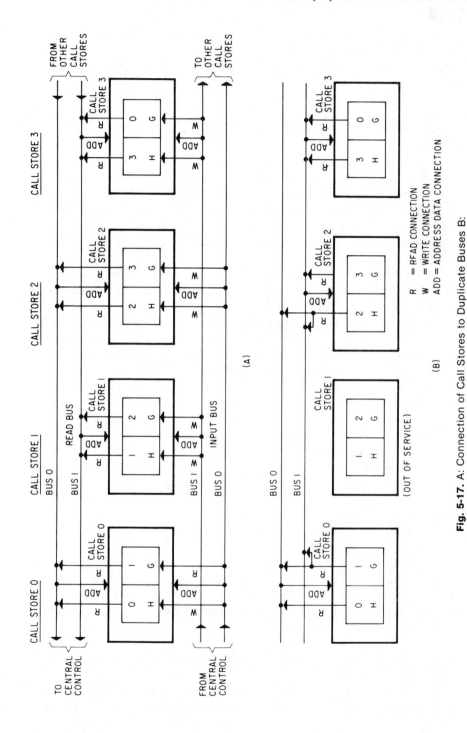

Fig. 5-17. A: Connection of Call Stores to Duplicate Buses B: Connection of Call Stores to Buses When Call Store 1 is Out of Service

As a result, the data read from this H half, likewise, are sent over both buses. This flexible arrangement for the call stores and their interconnecting buses is similar in many respects to that provided for program stores as previously explained.

Program and Call Store Operations In 1A Processor

Program and call stores in the 1A processor originally employed magnetic-core memories. But, unlike the memory systems described for the No. 1 ESS, the same basic ferrite-core memory module was employed for both program and call stores in the 1A processor. This memory module had a capacity of 65,536 binary words (26 bits per word), and was capable of operating in a cycle time of 1.4 microseconds for either *read* or *write* functions.

The ferrite cores were arranged in two planes or layers, mounted back-to-back in each module of program and call stores. Each ferrite core had a diameter of 0.023 inches (23 mils), and a layer contained an array of 832 by 512 cores. The arrangement was similar to that shown in Fig. 5-3 except that only two wires passed at right angles (orthogonal) through the center of each ferrite core. The horizontal wire was termed the *word* or *A Line,* and the vertical the *bit* or *B Line.* The *inhibit* and *read* wires, provided in the No. 1 ESS (see Figs. 5-3 and 5-14), were not required for the 1A processor.

The above-mentioned pair of orthogonal wires through every ferrite core performed *write* and *read* functions in addition to their normal core-selection operations. These roles were possible because of the two-wire, coincident-current technology that had been developed by the Bell Laboratories for this magnetic memory system. Figure 5-18 is an abridged schematic diagram of the principal elements of the initial magnetic memory system used in the 1A processor. The following brief explanations of its operations should acquaint the reader with the main precepts of the aforementioned two-wire, coincident-current technology. It is similarly utilized in the semiconductor memory systems.

The ferrite cores are arranged in pairs of an 8-by-8 matrix, one matrix in the front plane and the other in the back plane of the memory module. Each of the bit or B lines in the planes pass through the centers of 8 cores. One end of the bit lines in the front and back planes connects to respective terminals of a balun transformer. The other ends connect in pairs to the diode bridge rail that is associated with the particular bit-line select current source. These connections are shown in the simplified schematic diagram of Fig. 5-18.

The balun transformer assures that the currents flowing in the front and back planes are about equal to one-half of the input current. The bit-

Fig. 5-18. Memory Module in Program and Call Stores of 1A Processor

line select current is larger than the total memory currents in order that all diodes in the diode bridge rail would be forward-biased. In this way, a virtual ground condition will exist at the ends of the bit lines as a means of attenuating parasitic currents caused by the selection process.

The word or A lines in each layer of a matrix are looped through their respective ferrite cores at right angles to the bit lines. The same types of diode bridge rail and current supply circuits are employed for the word-line selection process as for the bit lines. In addition, a bipolar current also is provided in the vertical input circuit, shown in Fig. 5-18, for selecting a core in either the front or back plane of the matrix.

The interface between central control and the memory-access circuits of program and call stores is the *controller*. It generates all the timing pulses and controls the selecting circuits to pick a particular core in the designated memory matrix. The controller also governs the readout of data from the core for transmission to central control, and to write data received from central control.

The following example of the core-selection process refers to the core (memory cell) located on bit line 1 and word line 1, shown shaded in Fig. 5-18: The *controller* first activates the specific *read* and *bit line* current sources, associated with the desired matrix (determined from the address data received from central control). About 150 nanoseconds later, the required word-line select, horizontal select, and bipolar vertical input current sources are initiated. This delay is necessary because the core switching signal is also sensed on the bit line. Note that the bit-line currents on the front and back planes (linked with the balun transformer) are in opposite directions as indicated by the arrows in Fig. 5-18. Therefore, only the current in word line 1 will be in the proper direction to switch the ferrite core on bit line 1 in the front plane (see shaded core in Fig. 5-18).

A change in the selected core's magnetism, that is, switching from binary 0 state to binary 1 or viceversa, induces a voltage in the secondary winding of the balun transformer. It is designated "core output signal" in Fig. 5-18. This voltage is amplified and detected to ascertain whether a binary 1 or 0 was stored in the selected core. The results are relayed to central control in the similar fashion as previously explained for call stores in the No. 1 ESS.

The *read* and *write* functions and related operations of program and call stores in the 1A processor are performed in a similar manner as described for call stores in the No. 1 ESS. The reader, consequently, should refer to Figs. 5-12, 5-15, and 5-16, and their associated explanatory texts for further information applicable to the memory circuits of the 1A processor. Recall that program and call stores in the 1A processor are designed for a memory access speed of 1.4 microseconds as compared to 5.4 microseconds for the processor in the No. 1 ESS. Moreover, the 1A processor can execute basic instructions in 700 nanoseconds. At this ultra-high speed, over 1.4 million instructions can be completed in 1 second. In the same period, the No. 1 ESS can execute only about 180,000 commands.

Semiconductor Memory Circuits

Semiconductor memory circuits, utilizing integrated circuits, have been replacing magnetic core memories in program and call stores of the 1A processor since 1977. Substantial savings in equipment costs and floor space requirements, as well as lower power consumption, have resulted. The original integrated circuit chips had a capacity of 4K bits or 4096 binary digits. Note that $K = 2^{10}$ or 1024 when referring to binary digits in random access memory (RAM) circuits. Continued improvements in semiconductor technology, especially in the use of insulated-gate field-effect transistors (IGFET) have substantially increased the memory capacity of silicon integrated circuit chips.

For instance, the initial semiconductor memories employed flip-flop circuits that required several transistors in each memory cell. Subsequent developments by the Bell Laboratories reduced this to only one transistor per memory cell. This reduction and further advances in semiconductor fabrication technology made possible the production of single chips with the capacity of 16K bits initially. It has now been expanded to 64K or 65,536 bits. The new 64K RAM chips are contained in the same 16-pin dual-in-line package (DIP) originally used for the 16K chip. Furthermore, the memory access time has been lowered to 170 nanoseconds, thereby expenditing data processing in the 1A processor.

The above-mentioned semiconductor chips comprise a series of memory cells in a matrix of horizontal rows and vertical columns. The 16K RAM chip, for example, has 16,384 memory cells arranged in an array of 128 rows and 128 columns. A memory cell is situated at each intersection of a row and column. Consequently, there are 128 memory cells in each row on the 16K RAM chip.

The selection of a particular memory cell for write or read tasks is initiated by a unique combination of row and column address information in the form of binary digits (bits). This data is received by the respective *row* and *column address decoders* from buffer registers in central control (see Fig. 4-18) during memory operations. The row and column driver circuits then are activated to connect to the desired memory cell. The block diagrams in Fig. 5-19 illustrate the major components of the 16K RAM chip and the selection of the memory cell located at the intersection of row 100 and column 105.

Each memory cell consists of an IGFET (transistor) and a miniature capacitor as shown in Fig. 5-20. Information is stored in the memory cell as a charge on its capacitor. For instance, a binary 1 is written into the memory cell by charging the capacitor under control of its associated transistor. The memory cell is read by sensing the amount of the voltage charge on its capacitor. This is accomplished by the column-sensing amplifiers shown in Fig. 5-19.

However, the charge on a memory cell's capacitor changes rapidly because of current leakage. Hence, information can be retained for only about 4 milliseconds. It is necessary, consequently, that the capacitor's charge be periodically regenerated or "refreshed." This is normally required approximately 250 times per second to assure that the correct stored data is re-

Fig. 5-19. Major Elements of 16K RAM Chip (Courtesy of Bell Telephone Laboratories)

tained. To facilitate this regenerating or "refreshing" process, row decoder circuits are designed to "refresh" the charge in all memory cells in one row when the address of that row is specified by the associated row address decoder. Thus, the entire memory in a RAM chip can be regenerated by addressing all rows in sequence.

Fig. 5-20. Memory Cell in 16K RAM Chip (Courtesy of Bell Telephone Laboratories)

In this connection, a 16K RAM chip (which has 128 rows) would need 32,000 (250 x 128) "refreshing" operations per second to retain its stored data. In order to maintain electric compatibility with the 16K RAM, the new 64K RAM chip of Bell Laboratories is designed for 128 rows but with each row containing 256 memory cells. Consequently, the 32,000 per second "refreshing" operations likewise are applicable to the 64K RAM chip. Thus, these newer chips can provide equivalent electrical performance at much lower cost per bit to afford more economical electronic switching systems.

Questions

1. What two types and designations of memory systems in the No. 1 ESS are utilized by central control in processing calls? In what form are the instructions and information stored in these memory systems?
2. What is the basic element of the memory cell used in program and call stores in the No. 1 ESS? What are its main characteristics? What conductors intersect in the memory cell and what are their purposes?
3. What main function is performed by a memory cell in program store? In call store? What is the relationship between the direction of magnetization of a memory cell and its stored digit?
4. What additional memory systems are furnished in the 1A processor? Describe the new feature provided by these memory systems.
5. What important programs are stored on the disks and magnetic tapes in the 1A processor? What important function is served by the magnetic tapes?
6. How many bits comprise the binary words of program and call stores in the No. 1 ESS? In the 1A processor?
7. What operations are performed by call store? What three or more types of information are stored in the call store memory of the No. 1 ESS?
8. How many binary word locations are contained in each memory module of a call store in the No. 1 ESS? What is the total binary word capacity in one call store? How are these binary words divided?
9. What is the basic storage element in the memory module of call store in the No. 1 ESS? What function does it perform? What conductors pass through this element? Briefly describe their functions.
10. What happens to the stored information in a memory cell during the read operation in call store? Briefly describe the write operation in call store in the No. 1 ESS.

6
Stored Program Control and Call Handling Operations

Stored Program Organization

The stored program in the No. 1 ESS, No. 1A ESS, and similar electronic switching systems is designed to provide real-time operations of the switching system in response to random traffic, or varying customer calling rates. For instance, a customer expects to hear a dial tone immediately upon picking up his handset. Likewise, upon completion of dialing the customer expects to hear the ringing signal within a second or two. Random traffic refers to the fact that calls may be handled at an average rate of about 30 per second in the busy hour. During one second, however, no calls or only two or three may be processed, while during another second as many as 60 may be initiated.

An electronic switching system, consequently, must be capable of processing random traffic as quickly as possible in order to approach the real-time requirements for good service. To achieve this goal is the main function of the stored program control (SPC) system. In the No. 1A ESS, the SPC is designed to handle as many as 240,000 calls in the peak busy hour. It also has the capability of controlling new service features as well as future growth needs. In addition, the SPC arrangement provides for automatic error detection, diagnosis of trouble conditions, administrative and maintenance operations, automatic message accounting (AMA) functions, and other call-handling procedures.

The stored program part consists of instructions, translation information, and related data that are stored in program store. It also includes the latest changes concerning customer lines and trunks that are in the temporary memory of call store. Figure 6-1 portrays the main elements of the stored program control system of the No. 1A ESS. In essence, the stored

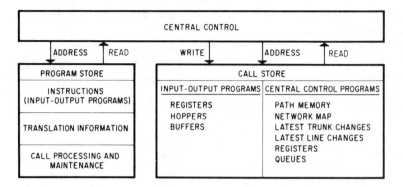

Fig. 6-1. Stored Program Control System Elements in No. 1A ESS.

program consists of approximately 200,000 instructions of which a little over half are devoted to call processing and related functions, with the balance covering automatic maintenance operations of the electronic central office. These instructions are read by central control one at a time from program store or from call store as needed. Note that neither the stored program nor central control by itself can carry out any functions. Rather, central control is the means of implementing the actions ordered by the stored program. For example, a function such as detecting a customer's line request for dial tone results from the combined effects of the program instructions read one at a time from program store and executed by central control.

Stored Program Control Analogy

In Chapter 2, the operations of electronic switching systems were explained by an analogy to the operations of the antiquated manual switchboard which was assumed to be run by electronic operators or automatons instead of human operators. This same analogy will be used to simplify the explanations of the operations of the stored program control system and its program organization. For this purpose, we will assume, in the case of the No. 1 ESS, that the switchboard is controlled by an electronic supervisor and an electronic operator.

The work performed by the electronic operator is not complicated but it is time consuming. It includes scanning the face of the switchboard for lamp signals indicating calls being originated, watching the cord lamps for a lighted lamp indicating disconnection of a call, obtaining the called number from the calling customer, and plugging and unplugging switchboard cords. These tasks may be classified as input-output actions or work performed in accordance with input-output programs. It is essential that these tasks be carried out at regular intervals to avoid loss of information and resultant down grading of service.

The electronic supervisor's tasks are more complicated because they involve actions such as analyzing the significance of digits received for various calls in progress, deciding what connections must be established or taken down at the switchboard, and recording the necessary information needed for billing of calls. These particular tasks, however, can be deferred for a short interval without affecting service.

To record the aforementioned types of information in the No. 1 ESS, the electronic supervisor and electronic operator employ separate electronic slates. The electronic slate used by the electronic supervisor corresponds to the call stores that are directly associated with central control in No. 1 ESS offices where signal processors are installed. Figures 2-5 and 4-14 illustrate the role of signal processors. Note that the electronic operator's slate is analogous to the call stores linked with the signal processors. These electronic supervisor–operator relationships are depicted in Fig. 6-2. Also, this electronic supervisor–operator analogy is applicable to the No. 1A ESS and other electronic offices equipped with the 1A processor. However, because of the higher processing speeds of the 1A processor, it is desirable to assume

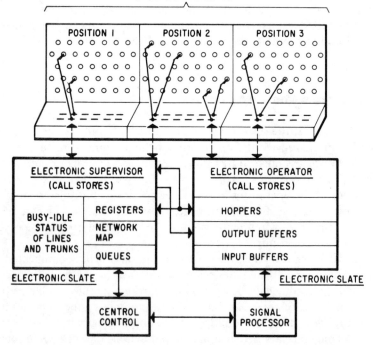

Fig. 6-2. Manual Switchboard Analogy for Stored Program Control in No. 1 ESS

that the electronic supervisor handles both its job and that of the electronic operator in the following analogous descriptions with respect to the No. 1A ESS.

In the No. 1 ESS analogy, the electronic operator communicates with the electronic supervisor by recording information in certain areas of its electronic slate called *hoppers*. For example, when an originating call is detected, the electronic operator enters this information in a list designated *service request hopper*. At a later interval, the electronic supervisor will unload or take out, one at a time, the data in this hopper in order to carry out the required action for each entry.

The electronic supervisor communicates with the electronic operator by recording data in areas of the latter's slate known as *output buffers*. For instance, when the electronic supervisor has decided what particular switchboard cords will be utilized to plug into specific jacks in handling a call, it will write the appropriate information into an available output buffer in the operator's electronic slate. The electronic operator, will subsequently unload this information from the output buffer and will then proceed to perform the required actions, such as inserting cords into specified jacks on the switchboard.

Other portions of the electronic supervisor's slate are used to record different types of information. For example, the busy-idle status of all switchboard cords (corresponding to the line-link and trunk-link network) is recorded in a certain part of the electronic slate (analogous to the network map in call stores). Other areas, called *registers,* are used to record data pertaining to the progress of calls. In this connection, the electronic supervisor selects and assigns a register to each call originated by a customer. This register is used by the electronic supervisor to accumulate the digits of the called number. These digits are received from the electronic operator via the digit hopper in which they are initially recorded, one at a time, as submitted by the calling customer.

In the course of handling a call, the electronic supervisor may find that certain facilities, such as a register, are not available because of traffic loads. In this event, it will record this information in an area of the electronic slate designated as the *queue* or waiting list for registers. This list or queue will be frequently checked by the electronic supervisor to ascertain whether it is possible to obtain a register to serve the waiting customer lines. The queue, in effect, serves as a memorandum to remind the electronic supervisor about tasks waiting to be accomplished.

Call Processing Analogy in No. 1 ESS

When signal processors are not provided, as in small and medium-size No. 1 ESS electronic offices, the electronic supervisor in our analogy also performs the electronic operator's tasks as previously stated. The electronic supervisor follows a prescribed schedule in carrying out the various call processing tasks. For instance, a particular task is initiated by checking a hopper or queue for any recorded data. Next, actions are performed which may involve recording information in a register or in an output buffer. The electronic supervisor proceeds from one item or entry to the next one until the bottom of the hopper or queue is reached. Then, it continues to the next hopper or queue in accordance with the list of tasks to be accomplished. When the end of the list is reached, the electronic supervisor starts over again.

Every 5 milliseconds, a clock circuit will signal the electronic supervisor to interrupt work and to record the data on the electronic slate. In this manner, it will be able to resume exactly where it left off when the clock signal was received. The electronic supervisor now undertakes a scheduled list of jobs which would have been accomplished by an assisting electronic operator if one were provided. Some of these tasks are performed during the interrupt signal every 5 milliseconds, while others are carried out every 10 or 25 milliseconds. The amount of time spent by the electronic supervisor on these secondary tasks depends upon the traffic load. Normally, the electronic supervisor would resume the main tasks after about 1.5 milliseconds.

Consequently, when a signal processor and its associated call stores are provided in the No. 1 ESS, the electronic supervisor can handle a greater volume of calls. For example, the electronic supervisor can continuously attend to the main functions of unloading hoppers and loading output buffers without interruptions. At the same time, the assisting electronic operator (utilizing the signal processor and its associated call stores) can follow the usual 5-millisecond schedule of loading hoppers and unloading output buffers.

Stored Program Functional Groups

The stored program, for purposes of call processing, consists of the following four functional groups with each group controlling a particular stage in the call processing procedure:

1. *Input-Output (I/O) Programs.* The input programs gather particular types of information for call processing purposes. For example: customer's dial pulses and digits received, line disconnections, and the idle or busy state of all lines, trunks, and service circuits. The output programs transmit data, obtained from the processing operations, to the switching network and other peripheral equipments. This output program data results in the operation and release of ferreed switches that provide interconnections in the switching network, and in the operation of trunk and service circuits.

2. *Operational or Call Processing Programs.* These programs are concerned with examining and processing the information pertaining to a particular type of call or to a phase of a call. These programs can also call on subroutines to collect data for use by the output programs. For example, the translation subroutine may be requested to supply data for the dial connection program.
3. *Service Routines.* Programs that process functions which are not solely related to one type of call or to a particular aspect of a call are known as service routines. An example is a program requesting a translation service routine to determine the specific line scanner equipment number that corresponds to a given directory number.
4. *Executive Control Program.* This is the main program which schedules the work of the input-output and call processing programs.

Each of the various programs in the stored program control system assumes and relinquishes direction of central control on a strict time schedule that is governed by the executive control program.

Stored Program Functions of Call Store

Call store is directly associated with the stored program control (SPC) system because its temporary memory acts as a clearing house for information between the above functional groups. Call store, for this purpose, is divided into sections or areas, each with a number of binary words or memory slots. These areas have registers, hoppers, and buffers in which information is deposited and withdrawn as required in processing a call.

For instance, one or more registers is associated with each call being processed. The hoppers receive information from the input programs and this information is then processed by the operational programs. In turn, the operational programs together with the subroutines send the resultant data to the buffers for use by the output programs. Moreover, the information in many areas of call store is constantly modified or updated as a call is processed. Data recorded by one program may again be used later by the same or other programs. Thus, the handling of any call involves a constant interaction between program store and call store with central control carrying out the resultant orders.

Each area in call store of the No. 1A ESS consists of one or more binary words of 24 bits each. In addition, 2 bits are added for error-checking purposes. All of the 24 bits may be utilized to store a particular information item or data. The binary word also may be divided into parts of one or more bits. For example, the entire 24 bits of a word may be used to store the identity of the trunk-line network termination of a service circuit connected for a specific call. Groups of 4 bits may be employed to store the three digits of the called office code. Likewise, groups of 4 bits in another binary word may be utilized to record the called line number. Figure 6-3 illustrates this use of such area in call store. The first and second binary words are used to store the called number by its office code and line digits of 555-1210. The third binary word identifies the trunk-link network termination as 342. The equivalent decimal numbers of the binary digits in the 4-bit groups are derived from the binary coded decimal table illustrated in Table 3-1.

BIT POSITION NUMBER

BINARY WORD	23		15	10	5	0	DECIMAL NUMBER
		GROUP 6	GROUP 5	GROUP 4	GROUP 3	GROUP 2	GROUP 1	
FIRST	1	0 0 0	0 0 0 0	0 0 0 0	0 1 0 1	0 1 0 1	0 1 0 1	555
SECOND	1	0 0 0	0 0 0 0	0 0 0 1	0 0 1 0	0 0 0 1	1 0 1 0	1210
THIRD	1	0 0 0	0 0 0 0	0 0 0 0	0 0 1 1	0 1 0 0	0 0 1 0	342
LAST BINARY WORD	1	0 1 0	1 0 1 0	0 1 1 0	0 1 1 1	1 0 1 0	0 0 0 1	206701

(BIT NO. 23 IS USED FOR PARITY CHECK PURPOSES IN NO. 1 ESS)

Fig. 6-3. Example of Stored Binary Digits in Call Store Memory Area

Program Priorities

Since call processing operates in real time, the stored program control (SPC) system must respond immediately to signals and data that occur at times that are not under its control. Central control, for instance, must keep up with the flow of information from customers and other central offices. To ensure dependable operations, it must respond quickly to errors detected by the many trouble detector circuits in the various equipment units. Otherwise, improper handling of calls and poor service may result. Consequently, it is necessary to establish an order of priorities in the system to ensure that operations of a *nondeferrable* type are carried out under controlled time schedules.

For example, detecting dial pulses is a nondeferrable task because pulses may be missed if this program is postponed for 5 milliseconds. Deferrable programs are those that can be deferred for 5 or more milliseconds without noticeable effects on service. Maintenance tasks may be either deferrable or nondeferrable depending upon the particular task. That is, if a call store unit becomes faulty, it is very important that it be immediately identified and removed from service. The diagnosis of a junctor link that has been removed from service, however, is a deferrable job that can be accomplished at other times.

A program being executed may be immediately stopped by an interrupt signal generated from one of several sources. In this case, a transfer is usually made to another program associated with the source of the interrupt signal. When the interrupt program completes its functions, the particular program that was interrupted resumes its functions as if it had not been interrupted. The interrupt signals enable central control to accomplish input-output operations that must be carried out at regular intervals, such as detecting dial pulses, scanning for line disconnections, and detecting ringing and answer signals. In addition, the interrupt signals cause central control to immediately take appropriate actions when trouble conditions are discovered, or if the execution of certain programs should be requested by mainenance personnel.

Program Interrupt Levels

Program interrupt operations may be divided into two major categories: interrupts due to the system clock which occur every 5 milliseconds in the No. 1 ESS (1.4 microseconds in the No. 1A ESS), and maintenance interrupts which may happen any time as the result of trouble detection. Program interrupts caused by maintenance have higher priority than those due to clock interrupts. For instance, if a maintenance interrupt occurs when central control is under the direction of the executive control program, central control will be transferred to the proper fault recognition program. In this manner, the defective unit can be identified and taken out of service. At the same time, the required information is entered in the maintenance register to permit the necessary diagnostic work to be accomplished later on a deferrable basis. The call processing work is then resumed. The interruption period in this case is normally less than 40 milliseconds in the No. 1 ESS.

The interrupt sources and the programs associated with them are classified into ten levels of importance. These interrupt levels are designated from the highest to the lowest order by the letters A to K, inclusive, except for the letter I. An interrupt source allotted to a specific level can only interrupt programs of lower levels, except that level B can interrupt level A. Most of the call processing programs in the No. 1 ESS are not related with any interrupt source. Consequently, they may be interrupted by any order of interrupt level. These programs are considered as operating at the lowest or base level L. Table 6-1, which illustrates the arrangement for these program interrupt levels for the No. 1 ESS, is also applicable to the No. 1A ESS.

The A level, which is the highest interrupt source, is initiated from the *master control center*. Levels B through G are interrupted by trouble detectors in the electronic switching equipment. Interrupt levels H and J are activated every 5 milliseconds when a signal processor is not installed in the No. 1 ESS. The J level initiates the input-output programs associated with central control. The H level is used to interrupt the J level's input-output programs whenever more than 5 milliseconds are required for their processing tasks. Interrupts of level K are generated only in No. 1 ESS offices equipped with signal processors when, for instance, a hopper overflows because of traffic loads. Base level L constitutes the *task dispenser programs* which are used to control specific hoppers or queues in call stores. Functions of the task dispenser programs are scheduled by the executive control program which is the main program.

Table 6-1. Priority Rank of Program Interrupt Levels

Interrupt Level		Interrupt Source
	A	Master control center
	B	Emergency action
	C	Central control
Maintenance	D	Call store
Programs	E	Program store
	F	Peripheral units
	G	Special tests
	H	Input-Output programs*
	J	Input-Output programs*
	K	Hopper overflow (used only when signal processors are installed)
	L	Base-level programs (directed by the executive control program)

*Interruptions of H and J levels occur every 5 milliseconds in No. 1 ESS

Executive Control Program

The executive control program schedules the work of the other functional program groups. For instance, every 200 milliseconds in the No. 1 ESS, the executive control program schedules the operation of the line scanning program in the input program group. This particular program directs central control to scan all customer line ferrods to ascertain which have changed state since the previous scan. Whenever a change is detected, the line scanning program temporarily stops and the equipment number of the calling line is written into the line service request hopper in call store. The required action on one line is completed before proceeding to scan the next line's ferrod. When all customer line ferrods have been scanned, the line scanning program returns control to the executive control program. At this time, any customer's line that has been off-hook since the last scan will be identified in the line service request hopper. These program functions are represented by the applicable block diagrams in Fig. 6-4 for the No. 1 ESS.

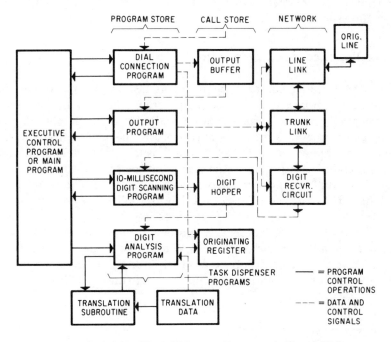

Fig. 6-4. Example of Task Dispenser Programs in No. 1A ESS.

The subsequent action to be taken is next decided by the executive control program. For instance, it could schedule an operational program, such as the dial connection program, to handle dial tone requests that are stored in the line service request hopper in call store. It may decide, instead, to allocate control to the line-link switching network or to another output program. In this case, a particular controller in the line-link network can operate only once in every 25 milliseconds. Therefore, the executive control or main program would direct other system operations in the interval.

Meanwhile, the service requests in the line service request hopper of call store are waiting for an operational program. Consequently, one step in processing a call may have no direct relation in time with the next step. Thus, as a specific call progresses through the electronic central office, control of its operation is handed back and forth between the main program and the operational programs within program store. Moreover, an operational program, for example, the ringing trip scanning program, may complete a processing action on one customer's line and then return control to the executive control program. Thus, the main program could direct the same operational program to the next waiting customer's line or it may initiate another operational program. In any event, a particular operational program is interrupted whenever a digit scanning program or network operation step is due.

Task Dispenser Programs in No. 1 ESS

The executive control program also schedules the operations of task dispenser programs. A task dispenser program takes data that was entered in a hopper or queue within call store, and refers it to an appropriate task program in program store for necessary processing. Consequently, task dispenser programs serve as links between the main program and the various operational or task programs that carry out the processing actions on the data in the buffers, hoppers, and queues in call store. Figure 6-4 shows a number of task dispenser programs that are utilized in program store. A brief synopsis of the operations involved follows.

Assume that a service request (customer's line off-hook) is received by the dial connection program from the line service request hopper (see Fig. 6-5). The executive control program first directs the dial connection program to load or temporarily store this request in an available output buffer

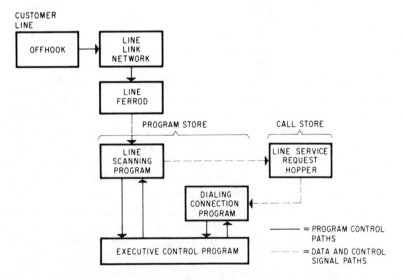

Fig. 6-5. Example of Executive Control Program Functions

in call store. Therefore, when scheduled by the main program, the output program will cause the output buffer to operate the appropriate controllers in the line-link and trunk-link networks. These actions will connect an idle dial pulse or digit receiving circuit to the originating customer's line. A dial tone will be sent at this time.

Every 10 milliseconds, all dial pulse or digit receiver circuits are scanned under control of the digit scanning program. All detected dial pulses are initially entered in the digit hopper in call store. The digit analysis program, as soon as scheduled by the main program, will direct the digit hopper to enter the recorded digits into an originating register in call store. Then, the digit analysis program with the aid of its translation subroutine and the translation data in program store will proceed to further process the call (see Fig. 6-4).

Task Priority Classes and Interject Programs in No. 1 ESS

The time required between the completion of a particular task dispenser program and its resumption depends upon the duration of all the intervening tasks. The buffers in call store, for instance, accumulate information of different degrees of importance and, consequently, must be unloaded in accordance with a precise schedule. The task dispenser programs, therefore, are grouped into five time-priority categories designated A, B, C, D, and E, and also an "interject" class which has the highest priority. Each priority class has a class control program that directs the operation of the task dispensers in that class.

Class A task dispenser programs in the No. 1 ESS, for example, cannot be delayed more than 200 milliseconds. Class E, the lowest priority, however, may be deferred as long as 3 seconds. Thus, the executive control or main program is arranged to utilize a frequency table, consisting of a continuously repeating sequence of the five classes, to determine the next priority class to be activated. Each priority class in the frequency table is assigned twice as often as the priority class below it:

A B A C A B A D A B A C A B A E A B A C A B A D A B

Note that the sequence contains Class A thirteen times, Class B seven times, Class C three times, Class D twice, and Class E once.

The interject class is not included in the above sequence, but interject programs may interrupt at any point. The executive control program makes a check for the existence of any interject requests in each task program. Such requests may be initiated by programs in interrupt levels H, J, or L as shown in Table 6-1. The interject requests are originated whenever a program in the interrupt levels encounters a task that cannot tolerate the delay of even a Class A priority (200 milliseconds). In such a circumstance, the executive control program interjects the particular task that must be performed between other task programs.

Translation Programs, Routines, and Subroutines

The translation programs in program store are used to convert known data, as supplied by a task dispensing program, into related data that is referred back to the requesting task program. The known data may be a three-digit office code, a four-digit directory number, or the identity of a trunk group or a network termination. A translation program that is common to a number of task programs is usually designated a *translation routine*. When a particular translation program or routine always returns control to its task program upon completion of its work, it is called a *translation subroutine*.

An example of the use of translation programs is illustrated in Fig. 6-4. Assume that the digit analysis program receives office code 555 from the digit hopper in call store. The digit analysis program will use this known data, 555, as an address to gain access to the related information in the translation subroutine. Translation data in program store are grouped into tables according to the information used to obtain access to them. Thus, the related translation information received by the digit analysis program for office code 555, would be loaded into an assigned originating register for processing.

Changes in translation data are sometimes required because of equipment reassignments or changes in service resulting from customer requests. Such modifications are not made immediately in program store, but are entered in the call store memory and designated as *recent changes*. In the No. 1 ESS, these recent changes are collected and then written on the aluminum memory cards in program store by mean of the memory card writer as explained in Chapter 5. Such modifications in the No. 1A ESS usually are initiated at the master control console by maintenance personnel. The required changes are made to the pertinent memory stored in the file stores (disk files) of the 1A processor. Subsequently, the resultant modified data is automatically entered into program stores. Translation service routines usually consult the recent changes entered in call store before referring to the translation data in program store.

Service Routines

During call processing operations, it is often necessary to initiate changes in the switching network and in the state of junctor, service, or trunk circuits. To accomplish these jobs, the task program calls upon appropriate service routines. For instance, the ringing connection program may request a network service routine to search for an idle ringing circuit. It may also request paths through the line-link and trunk-link networks to connect the ringing circuit to the called customer's line.

An example of some functions of a network service routine during an intraoffice call is shown in Fig. 6-6. The ringing connection program, in this example, calls upon a particular network service routine designated the network path hunt subroutine. It requests this service routine for the related data needed to select a ringing circuit for connection to the called customer's line, and for an audible ring circuit that will be connected to the originating customer's line. The network path hunt subroutine first refers to the path memory and network map in call store for the related data on available audible ring and ringing circuits, and for the idle paths needed to connect these circuits to the network terminations involved. The selected paths are marked busy in the path memory and network map in call store.

The network service routine next loads an output buffer in call store, usually a peripheral order buffer (POB), with the necessary information for performing the required interconnections. The output buffer, as soon as activated by its associated task dispenser output program, will send the necessary control signals to make the required interconnections in the line-link and trunk-line networks as indicated in Fig. 6-6. Other service routines are used in a similar manner to place orders in peripheral order buffers (POB) for operating scanners, signal distributors, and central pulse distributors in connection with call processing.

Network Map

A network map (see Fig. 6-7) is associated with the path memory in call store. The data in the network map provide the particular line links and trunk links that are available to establish a path through the network as well as those lines already busy. Moreover, a number of busy links may be combined into paths through the network system. Consequently, the network map itself cannot indicate at any time how lines, junctors, and trunks are associated by the existing paths through the network. It is therefore necessary to obtain this information by other areas in call store which collectively are called the *path memory*.

The record of the idle or busy state of all links or paths in an electronic central office is contained in the network map. The idle state is denoted by the binary digit 1, and the busy condition by binary digit 0. Thus, call store and not the switching network must be consulted by central control in order to obtain idle paths in handling a call. A simplified diagram of a network map appears in Fig. 6-7. It shows the possible paths that may be listed in the network map of call store for an intraoffice call from customer A to B. Re-

Fig. 6-6. Example of Translation and Network Service Routine Program Functions

184

call that a binary 1 denotes an idle link and a binary 0 implies a busy link. Therefore, only two idle paths would be available between customers A and B through the line-link network in this example as indicated by the respective arrows on the idle links that are marked 1.

Functions of Path Memory on Intraoffice Calls

The path memory in call store consists primarily of three types of binary words. One type is permanently associated with each output termination of the trunk-link network. Another is permanently associated with each junctor termination on the line-link network. And several types are assigned as required to the trunk-to-trunk junctors used for intraoffice calls.

Whenever a customer's line is connected through the line-link network to a junctor, the particular path memory's line word associated with the

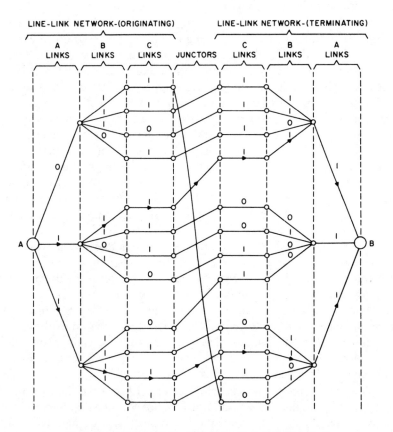

Fig. 6-7. Network Map in Call Store for Intraoffice Call

junctor will store the identity of the customer's line. The path memory's trunk word that is associated with the trunk termination in a line-to-trunk connection will contain the identity of the particular junctor involved. In a trunk-to-trunk connection, each of the two path memory trunk words concerned contains the identity of the pair of trunk-to-trunk memory words assigned to the trunk-to-trunk junctor. The trunk-to-trunk memory words in this type of call connection also contain the identity of the two trunks involved and of the trunk-to-trunk junctor.

Figure 6-8 shows the various programs and routines in program store, the related call store hoppers, buffers, registers, and the network equipment with the associated circuits that are usually employed during the ringing and talking connection phases of an intraoffice call. The ringing circuits are scanned every 100 milliseconds, under the control of the ringing trip scanning program, to ascertain whether any called lines have answered. This scanning is accomplished in a manner similar to that described for the customer's line scanning program.

A called-line answer or off-hook indication is first detected by the ringing circuit. This causes the ringing trip scanning program to enter the scanner address data in the ringing trip hopper in call store. When the executive control program schedules the talking connection program to function, each entry in the ringing trip hopper is referred to this program which also may be considered as a ringing answer detection program. The talking connection program then utilizes the path memory idle trunk list to determine the identity of the network appearance of the ringing circuit and of the associated ringing register. Upon receipt of this information, concerning the particular network path involved, the talking connection program proceeds to load this data to an output buffer in call store, designated a peripheral order buffer (POB). The high-priority output program immediately directs this POB to release the ringing register and the related audible ring and ringing circuits.

The talking connection program next makes idle the network paths that were previously used by the audible ring and ringing circuits. It then loads the output buffer or POB with orders to establish the previously reserved talking paths in the line-link network between the originating and called lines. The talking path, which is designated AB and CD in Fig. 6-8, utilizes an associated junctor circuit for operational and control purposes. Note that registers are not associated with intraoffice calls in the talking phase. However, all junctor circuits are scanned every 100 milliseconds under control of the junctor scanning program.

The path memory also provides essential reference data during the ringing stage of an intraoffice call. For instance, assume that the originating customer abandons the call while the called line is being rung. The scanner address of the audible ring circuit would be in the path memory. This infor-

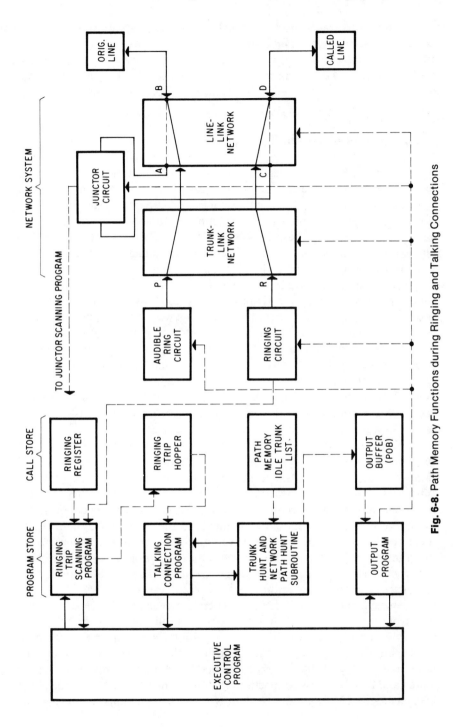

Fig. 6-8. Path Memory Functions during Ringing and Talking Connections

mation would be translated by a translation subroutine into the identity of the related ringing register and the particular trunk-link terminations of the associated audible ring and ringing circuits. These network terminations, such as those marked P and R in Fig. 6-8, would be subsequently released by timely orders issued by the appropriate output program.

Program Timetables and Timing Functions

It is seen, from the foregoing, that every action in the process of making a connection between the calling and called lines is controlled by one or more stored programs. The executive control program schedules these functional stored programs in a manner analogous to an orchestra leader who cues in each section of the orchestra at the proper moment. The successful and efficient completion of calls depends on the precise scheduling of the various task programs, especially the input programs.

Note that input signals originate at customers' telephones or in other central offices, while output signals are controlled by specifically assigned programs. Thus, if any input signals are missed or improperly received, calls may go astray or be disconnected. Moreover, many input signals, such as dial pulses, exist only momentarily so that the program assigned to detect them must be scheduled precisely to the millisecond. Input programs, therefore, have precedence over other tasks. Consequently, in the No. 1 ESS, every 5 milliseconds the particular program in progress is halted by the executive control program to permit an input program, such as line scanning, trunk scanning, or junctor scanning, to take control.

Thus, in the No. 1 ESS, it is required that the transition from a call processing routine program to an input detection program occur every 5 milliseconds. It is not feasible, however, for the various task programs to time their own control periods. To do so would necessitate that the programs be written to account for both the random occurrences of inputs and the very accurate timing requirements of the peripheral system equipment. For this reason, the executive control program determines which program will have control at any given time. The programs themselves are not concerned with the timing functions since the 5-millisecond timing pulses are generated by a timing circuit in central control.

The specific timing indicators generally needed by the executive control program are provided by two main timetables in call store. They are designated low priority and high priority with respect to their associated input-output programs. Some examples of the related programs in these timetables are listed in Table 6-2. Each timetable actually is a matrix of binary words and digits. In this connection, the low-priority timetable in the No. 1 ESS has a matrix of twenty rows and twenty-three columns. Each row corresponds to a 5-millisecond interval so that the low-priority timetable covers a 100-millisecond period. The high-priority timetable has a matrix of twenty-four rows and twenty-three columns and, consequently, has a 120-millisecond cycle. Each column in both timetables is associated with a particular input-output program, for example, digit scanning, ringing trip scanning, and the ringing circuit answer programs.

The particular program is usually executed when it becomes due. However, it may be rescheduled at a lower priority. To accomplish this, the program would write a binary 1, which is called a *flag bit*, in certain control registers in call store. This flag bit causes the executive control program to schedule the operation of specific task programs at the desired intervals.

Table 6-2. Programs with High and Low Priority in Call Store

Programs	Operating Sequence (milliseconds)
High-priority	
Digit scanning	10
Abandon and interdigital time-out	120
Call charge monitor tape output	5
Low-priority	
Peripheral order buffer execution	25
Ringing trip scanning	100
Outgoing trunk scanning	100
Junctor scanning	100

The high-priority timetable is illustrated in Fig. 6-9. Each row in this timetable matrix corresponds to a binary word and is associated with a particular 5-millisecond interval within the 120-millisecond period of this timetable. The rows in the timetable are read out in turn, one every 5 milliseconds. Whenever a binary 1 appears, the assigned program for the intersecting column is executed. If a binary 0 appears, the corresponding program is omitted during that interval. The readout process is continuous so that when all rows are read out, the cycle starts all over again.

The use of the high-priority timetable, per Fig. 6-9, may be explained by the following examples. Assume that the digit scanning program is assigned to vertical column 8. Since this particular program is executed every 10 milliseconds (see Table 6-2), column 8 will have a binary 1 in every other row as shown in Fig. 6-9.

For another example, assume that the abandon and interdigital time-out program, which is accomplished every 120 milliseconds (per Table 6-2), is assigned to column 15. It will be executed at every row 10 because a binary 1 appears only at the intersection of this specific row and column 15. Moreover, a program to be performed every 20 milliseconds would have a

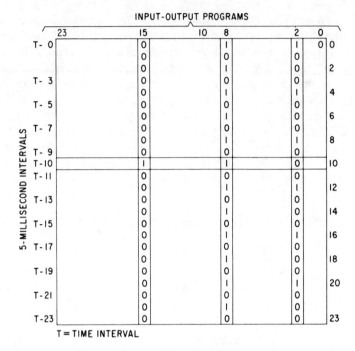

Fig. 6-9. Matrix of High-Priority Timetable

binary 1 in row numbers 0, 4, 8, 12, 16, and 20 as indicated in column 2 of Fig. 6-9.

It is apparent that these timetables cannot be used for sequences that do not repeat themselves either every 100 or 120 milliseconds. It is necessary, therefore, to utilize a revolving type of counter for other regular repeating time sequences. This type of counter consists of a number of bits in a call store word. The bit at the extreme left side is rotated to the opposite end of the word at regular intervals. Whenever that extreme left bit is a binary 1, the assigned program is performed. The control signal to rotate the extreme left bit is taken from the high-priority timetable.

As an example of this timing sequence, assume that a certain program is executed repeatedly at 45 and 60 milliseconds. This specific sequence may be triggered by a 7-bit revolving counter as represented in Fig. 6-10. Thus, at every third interval of the high-priority timetable (every 15 milliseconds), its output would be applied to this 7-bit counter in order to rotate the leftmost bit. A complete clockwise rotation of the counter from position 0 to position 6 requires 105 milliseconds. Consequently, a binary 1 will appear at the leftmost position (6) every 45 and 60 milliseconds to initiate the execution of the program.

CALL STORE WORD

BIT POSITION	6	5	4	3	2	I	0
BINARY DIGIT	I	0	0	0	I	0	0

Fig. 6-10. Example of Revolving Counter Utilizing Bits in Call Store Word for 105-Millisecond Cycle

Call Store Register Operations with Stored Programs

The stored program used in the No. 1 ESS, No. 1A ESS, and similar electronic switching systems is generic. A single program, therefore, can be utilized to serve many electronic central offices, some having different features or characteristics. As a result, any address, quantity, or option that can vary from office to office must be referred to a specified area of program store, or to call store in the case of recent changes. The job of the task dispenser programs is to examine the data stored in the buffers, and then to decide what particular task programs should be called in to perform the required operations; for example, to direct the scanning of specific customers' lines, or to initiate the ringing connection program.

When this work is finished, the task program involved returns control to the task dispensing program which checks for any waiting interject orders. If none is found, the task dispenser program again examines its associated buffer and the cycle continues. When all task dispensers in one priority class, such as Class B in the No. 1 ESS, are completed and their buffers are cleared, the task sequence moves to the next lower priority category, Class C in this case.

All the individual parts of a task program are tied together in the call store memory. Each call in progress is assigned to a call store register which consists of memory areas for the temporary storage of input and control data. Some phases of a call necessitate more memory space than other stages. For instance, seven or ten digits may be stored during digit scanning, depending on whether the call is local or toll. Consequently, there are different-sized registers for the various stages of a call. The program subroutines, however, handle all call control registers in the same manner because these registers utilize a standard format.

Figure 6-11 is a simplified illustration of the layout of a typical call store register in the No. 1 ESS. The first four words (each word consists of 23 bits) of any call control register are used for specific storage purposes. The first word is called the *state word* because it identifies the register by its call state and by its function—originating, disconnecting, outpulsing, etc. For instance, when the register receives an input from a task dispenser, the state word selects a group of task programs appropriate to the particular

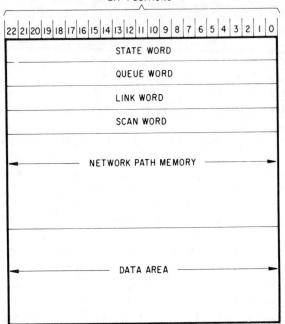

Fig. 6-11. Simplified Layout of Call Control Register in Call Store of No. 1 ESS

processing stage. The register then selects the exact program by referring to the type of task dispenser program that delivered the input data.

The second word is termed the *queue* because it is used to insert the register on a waiting or timing list. This is often necessary because peripheral equipment, such as an audible ring circuit, or a ringing circuit, may be busy on another call. In this event, when the required ringing circuit becomes available, the queue control program will assign it to the specific register on the queue word. Then, the queue control program will transfer control to the task program that was selected by the state word.

The third word is designated the *link word*. It is used to hold another call control register's address if more than one register is associated with the call. Any number of registers may be linked together by placing the address of the state word of the second register in the link word of the first register, the address of the state word of the third register in the link word of the second register, etc. This chain is closed by putting the state word of the first register in the link word of the last register. Buffer entries to any one register in the chain can be passed to all of them. When a register receives a buffer entry, its state is changed so that future entries will select the task program appropriate to the current state of the call in progress.

The fourth word is the *scan word* which is not always used. Its purpose is to link a call store register to a scanning register when the call control program has requested the scanning of some particular points in the system. The next group of words in the register is set aside for the network path memory. It may differ in length of words from one type of call control register to another. The storage area following the path memory is the *data area*. Its size depends upon the amount of data required by the call control programs.

It is obvious that the stored program is the operating intelligence of the electronic switching system. The program structure is determined to a large extent by the call handling requirements of the electronic switching system, and the fact that it must respond to service requests as they occur, that is, in real time. Specific data such as equipment quantities and translations, which vary from office to office, are not written into the generic program. However, they can be easily added to the memory system in any office. The program can be readily modified to incorporate new service features or changes. Consequently, it may be asserted that the No. 1 ESS, No. 1A ESS, No. 4 ESS, and related electronic switching systems can be arranged to control telephone services and features that have not yet been foreseen.

SPC Operations in 1A Processor

The 1A processor is an improved and more sophisticated machine than the processor in the No. 1 ESS. However, the foregoing description of SPC operations in the No. 1 ESS are generally applicable to the call-processing programs and related system software of the 1A processor. To ensure maximum dependable operations, software is included in the 1A processor to permit continuous automatic self-testing of subsystems and major hardware and software elements.

This system software may be classified into general categories of *maintenance control, administrative*, and *utility programs*. The block diagrams of Fig. 6-12 show this division, and the following is a succinct summary of the main software functions.

Maintenance control programs allocate processing time and memory space to various programs that can automatically detect and solve both software and hardware (equipment) problems. For instance, if a hardware error is detected, the faulty equipment unit is identified by the maintenance program and removed from service. At the same time, the remaining equipment items in the particular subsystem are arranged to continue their call-processing functions. In most cases, recovery of normal operations occurs without interfering with the handling of calls.

Fig. 6-12. SPC System Software of 1A Processor (Courtesy of Bell Telephone Laboratories)

If an operational spare is not available to replace a defective unit, the *recovery program* is activated to execute a "bootstrap" strategy. In this situation, all units in the particular subsystem in which the failure occurred are tested by the *recovery program*. The equipment units that pass the tests are arranged to work in the subsystem.

Maintenance programs also diagnose a program or call store that had been removed from service. These programs can accurately locate the trouble to within a few circuit packs. Thus, the defective circuit pack can be quickly replaced and the equipment item put back into service.

Software errors, likewise, are detected by routine maintenance programs. Faults caused by mutilation of the data base or part of a program are handled differently than hardware troubles. *Audit programs* are set up to analzye the data base and to restore the affected area on a time-shared basis with normal call-processing operations. In case the program is severely damaged, it is automatically restored from the disk memory in file stores. In this event, call-processing work is delayed until the damaged data has been corrected or restored by the transfer of corresponding data from file stores.

Administrative programs normally serve as interfaces between other programs (software) and the following equipment (hardware) of the 1A processor: file store, tape units, input/output devices, and the master control console (see Fig. 2-17). Some examples of the functions performed by various administrative programs are: transfer of data to and from input/output devices, file stores, and tape units in the 1A processor; collecting data on failures as they occur for later analysis, and the automatic reaction of the system to these failures; controlling the transfer into program store of programs from the disk memory (file store) for subsequent execution; and to temporarily allocate space in program or call store for the storage and execution of programs that are not an integral part of the system software.

Another example is the *master control console administrative program*. It interfaces with the master control console (MCC) and the processor peripheral interface (PPI) unit as illustrated in Fig. 2-17. One of its important functions is to read the status of keys and lamps on the MCC for other programs. This function is accomplished by a matrix of scanners and signal distributor points that connect to the MCC. This program also keeps a record of the matrix for use as a backup in case of hardware failure.

Utility programs primarily assist maintenance personnel in testing initial and updated programs and to debug faulty software features. Control of utility programs, therefore, is associated with maintenance control programs as indicated in Fig. 6-12. In this connection, the 1A processor has many programs and a large data base. Moreover, new features are constantly being added to its programs. Thus, there are bound to be some latent errors in the system software. Consequently, utility programs are arranged to have full access to data in the system so that they can sample essential data and analyze system behavior.

For instance, maintenance personnel may desire to obtain information why a specific program is not performing properly. Likewise, information concerning the selection of a particular trunk group in an office, or reports on the way in which known data is accessed or changed, may point to the solution of a problem.

The above summary of the major functions of the system software in the 1A processor affords a brief overview of SPC operations in the No. 1A ESS. Recall that call-processing programs of the No. 1 ESS and the 1A processor are compatible. The reader, accordingly, is referred to the preceding pages for more detailed descriptions of SPC system operations that are applicable to both types of electronic switching offices.

Questions

1. What major stored program elements are contained in program store? Approximately how many instructions comprise the stored program and what percentage is devoted to call processing and related functions in No. 1A ESS?
2. Name the four functional groups of the stored program and briefly describe their functions. Which one is the main program?
3. What three elements in call store are associated with stored program control operations? Briefly describe their functions.

4. What are the two major program interrupt categories? Which has the higher priority? What is the number of interrupt levels and how are they designated? Which one is the lowest level and what function does it perform?

5. What are task dispenser programs in the No. 1 ESS? and how are they scheduled?

6. What functions are performed by a translation program? What is the designation of a translation program that always returns control to its task program?

7. Where is the network map located and what does it contain? What indicates the state of a path in this map?

8. Describe the main elements of the path memory in call store.

9. What major timing indicators are provided for the executive control program in the No. 1 ESS? What are their designations? Briefly describe their structure and timing periods.

10. What are the main system software categories in the 1A processor? What major function is performed by the recovery program?

7
Switching Networks, Master Control Center, and Power System

Switching Network Precepts

The step-by-step, panel, rotary, and crossbar electromechanical systems, briefly described in Chapter 1, utilize various mechanical and electromechanical devices to accomplish their switching functions. For instance, in step-by-step central offices, the selector and connector switches themselves make the required interconnections between the calling and called lines in sequence with the dialed digits. In panel offices, the switching operations were performed by selector brushes which glide over terminals on vertical banks under control of a register-sender. In crossbar systems, the marker first checks the S (sleeve) leads to find a set of idle switching links and an idle trunk. It then closes the corresponding crosspoints on selected crossbar switches to accomplish the network connections.

In the original design for an electronic switching system, the use of electronic means for the actual switching functions was also considered. In fact, solid-state devices and gas vacuum tubes were employed as the switching network in some of the first laboratory and trial installations. There were, however, some serious drawbacks. For instance, gas tubes and transistors were unable to handle the high amplitude 20-Hz ringing signals.

Solid-state devices such as diodes, transistors, and integrated circuits, likewise have deficiencies when used for switching telephone lines and trunks. These electronic devices can readily pass current in the forward direction and present a high resistance to reverse current flow. However, they can also rectify voice-frequency signals and, therefore, may impair transmission by introducing noise in other parallel paths. This condition is illustrated in Fig. 7-1. It shows six telephone line circuits in a central office, any one of which may be connected or switched to the outgoing trunk circuit,

Fig. 7-1 Proposed Solid-State Devices for Network Switching

To, Ro. This switching may be accomplished by applying a suitable bias current to the respective pair of switching diodes associated with the particular line circuit.

For example, by applying the correct bias to the switching diodes of line circuit 4, only this line will be connected to the outgoing trunk. The diode outputs of the other lines, however, are in parallel with each other as well as bridged to line 4. Thus, noise voltages may be introduced into line 4 and its connected outgoing trunk circuit, thereby downgrading voice transmissions. This condition would be aggravated as the number of parallel paths is increased.

Moreover, field trials and laboratory tests indicated the need of the physical opening of unused parallel paths in order to ensure good transmission, and to provide for compatibility with existing telephone plant and equipment. Consequently, mechanical devices, such as relay contacts or crosspoints on crossbar switches, were used rather than electronic means. A high-speed reed switch with balanced metallic contacts was originally developed by the Bell Telephone Laboratories to perform the switching network interconnections in electronic offices. It was designated the *ferreed switch.*

Ferreed Switch

The *ferreed switch* was the basic interconnection device initially employed in the switching network of the first No. 1 ESS and similar electronic central offices. It was soon replaced by the *remreed* switch, which is similar in most respects to the *ferreed* type. The composition and operation of both switching devices are described in order to effectively explain the interconnection functions of the switching network in electronic central offices.

Both the *ferreed* and *remreed* are two-wire magnetically latched devices. They consist of two miniature magnetic-reed switches, each sealed in a glass tube. In the ferreed, the two sealed-reed switches are mounted between two rectangular *remendur* plates. One reed switch controls the tip conductor and the other controls the ring conductor of the two-wire transmission path through the switching network.

Remendur is a two-state magnetic alloy which can be changed very quickly from one magnetic state to the other by means of very short current pulses through a surrounding wire coil. This particular magnetic material also has a high remanence, that is, it will stay magnetized until another current pulse switches it back to its previous state. This switching action was accomplished in a millisecond or less. The ferreed, therefore, was many times faster than other electromechanical switching devices.

Two types of ferreed switches were provided for the switching network in initial No. 1 ESS offices. They were termed the *crosspoint ferreed* and the *bipolar ferreed*. The crosspoint ferreed, or ferreed as it was called, supplied the crosspoint contacts for the T and R conductors in the eight stages of switching provided in paths through the line-link and trunk-link networks. The bipolar ferreed was utilized for two main functions. It disconnected a customer's line circuit from its associated scanner ferrod (see Fig. 2-9). This function is similar to that of the cutoff (CO) relay in the step-by-step and panel central offices. The bipolar ferreed switch was also associated with junctor switching frames. It provided access to circuits used to test a line or trunk for false cross and ground (FCG) conditions, and for no-test and restore-verify requirements.

The original ferreeds were assembled in a square array of 64 switches consisting of eight columns and eight rows. Figure 7-2 is a photograph of the basic 8 × 8 two-wire crosspoint array of ferreed switches used in the No. 1 ESS. The reed contacts were connected to horizontal and vertical straps that formed the network multiple. When connected by the reed contacts, these horizontal and vertical multiples furnished the T and R paths through the ferreed switch. The remreed switch is similarly organized.

The reed contacts of all ferreed and remreed switches always open and close under "dry" circuit conditions. That is, *no battery potential* is applied to the line, path, or trunk during the switching interval. Thus, the reed contacts do not open or close current-carrying conductors. Integrated circuits in the associated line scanner, trunk, junctor, and service circuits function (before and after the switching interval) to interrupt and apply, respectively, battery potential to the T and R conductors. This procedure ensures a very long life for these miniature magnetic-reed switches.

Fig. 7-2. Eight-by-Eight Crosspoint Ferreed Switch Assembly Initially Employed in No. 1 ESS (Courtesy of Bell Telephone Laboratories)

Operation of Ferreed Crosspoint Switch

The main components and the construction of the crosspoint ferreed are illustrated in Fig. 7-3. The two remendur plates are divided magnetically into two halves by the horizontal steel shunt plate which is positioned at their centers. The two control windings are wound on a form between the two remendur plates. Each control winding has about sixty turns, with forty turns on one-half of the ferreed assembly opposing twenty turns on the other half of the switch. This arrangement is represented by windings 1 and 2 in Fig. 7-3.

When only one winding is energized by a current pulse, the two halves of the remendur plates are magnetized in *series-opposition* (the adjacent ends have the same magnetic polarity). As a result, there will be very little magnetic flux through the reeds and their normal tension will open the contacts. This condition is portrayed in the right-hand sketch in Fig. 7-3. Now if both control windings are pulsed simultaneously with equal currents, both halves of the remendur plates will be magnetized (*series-aiding*) and the reed contacts will close. The left-hand sketch in Fig. 7-3 illustrates this state. The remendur plates retain their magnetic polarity after the current pulses cease so that the reed contacts remain closed. Note that ferreeds and remreeds do not require power to keep their contacts closed.

Fig. 7-3. Components of Crosspoint Ferreed Switch Initially Used in No. 1 ESS

Operation of Bipolar Ferreed Switch

The bipolar ferreed is constructed similarly to the crosspoint ferreed except that the two sealed reed switches are mounted between two magnetic alloy rods. One rod is a permanent magnet and the other is a semipermanent magnet made of remendur. The magnetic polarity of the remendur rod is controlled by the direction of the current pulse sent through its winding.

The sketches in Fig. 7-4 illustrate the construction of the bipolar ferreed switch. When a current pulse is sent through the control winding in the BA direction (left-hand sketch) the generated magnetic flux will add to the flux from the permanent magnet rod, and both will pass through the reeds. The reed contacts, consequently, will close because of the greatly increased magnetic flux.

When the control winding is pulsed in the opposite or AB direction (right-hand sketch), the resultant magnetic polarity of the semipermanent

Fig. 7-4. Components of Bipolar Ferreed Switch Initially Used in No. 1 ESS

magnet will attract the flux from the permanent magnet. As a result, almost all magnetic flux will bypass the reeds in the sealed glass tubes. The normal tension of the reeds, therefore, will cause their contacts to open. In either case, when the current pulses cease, the semipermanent remendur rod retains the resulting magnetic state. Thus, the bipolar ferreed remains operated or released after the current pulse.

Improved Remreed Switch

The remreed is the improved magnetic reed switching device developed by the Bell Telephone Laboratories to replace the ferreed switch in electronic switching offices. It utilizes all electronic controls, has greater reliability, and achieves a 4 to 1 space reduction over the former ferreed switching networks. The remreed switch, with its associated diodes, transistors, and silicon integrated circuits (IC), replaces the ferreed switches, latching relays, and discrete solid-state logic devices used in the ferreed networks.

The remreed switch is similar in many respects to the ferreed. It has two sets of overlapping magnetic reeds. Each pair is hermetically sealed in a glass envelope analogous to the ferreed. The reeds are made of remendur which is a semipermanent magnetic alloy as previously explained. The remendur reeds are magnetized by current pulses sent through the control windings so that they will hold or latch the sealed contacts in either a closed or open state. In contrast, the ferreed switch requires external remendur plates to provide the latching action. The remreed switch, consequently, provides a magnetic field that is closer to the reed contacts and results in more efficient operations. Also, magnetic interaction between adjacent remreed crosspoints is substantially reduced as compared to ferreed crosspoints.

Remreed switches are usually arranged in two 8 by 8 arrays. This assembly contains 128 crosspoints compared to the 64 crosspoints provided by an equivalent array of ferreed switches. The sealed reed contacts of the remreed crosspoints are held between two printed circuit boards which also contain the diodes and transistors for the electronic control of the remreeds. Figure 7-5 is a photograph of a typical remreed switch package or assembly.

Fig. 7-5. Remreed Switch Assembly (Courtesy of Bell Telephone Laboratories)

Figure 7-6 shows the major components of a remreed and a schematic drawing of its control windings. The remreed crosspoints are divided into upper and lower halves by a steel shunt plate similar to the construction of the ferreed switch. The magnetization of each half is controlled by a primary and a secondary differential winding; one is in the horizontal pulse path and the other in the vertical pulse path as depicted in Fig. 7-6. These two differential windings are wound on a form which surrounds the glass envelopes containing the reed contacts.

When a current pulse is sent only momentarily through the vertical pulse path, $V_1 V_2$, magnetic fields are produced in opposite directions above and below the steel shunt plate. These magnetic fields will magnetize the upper and lower parts of the sealed reed contacts in opposite directions, thus the contacts will release. For the same reason, a momentary current pulse only through the horizontal pulse path, $H_1 H_2$, likewise will cause the reed contacts to release. When current pulses are sent through both vertical and horizontal pulse paths simultaneously, a magnetic field will be produced in the same direction both above and below the steel shunt plate, and the reed contacts will close.

Fig. 7-6. Details of Remreed Crosspoint

The crosspoints or contacts of the remreed switch are associated in pairs, as in the case of the ferreed. One set of contacts is used for the tip (T) conductor and the other set for the ring (R) conductor of the two-wire line, trunk, or other network transmission path. The remreed switch, as the ferreed, employs the destructive current mark principle. That is, a current pulse passing through a particular column and row of remreed crosspoints will operate only the specific crosspoint at the intersection. All other crosspoints along the path will release. The operated crosspoint connects the T and R leads associated with the column to the two-wire pair associated with the row. At the same time, crosspoints in other rows and columns, which may have been left connected from previous calls, will be released. Thus, when a call is disconnected or abandoned, the path memory in call store is marked so that the links that were used by that call are idle. This destructive mark ensures that these links will be disconnected from their previously established path when needed for a succeeding call.

Switching Network Elements

The switching network is the octal type consisting of eight stages of remreed switches. The crosspoints of these *remreeds* establish two-wire metallic paths, commonly designated the tip (T) and ring (R) conductors, for the transmission of voice and control signals. This switching network, besides interconnecting customer lines and trunks, also connects lines to junctor circuits and to various service circuits (dial pulse receivers, ringing and tone circuits, and signal transmitters) during the processing of a call.

The main equipment units of the switching network consist of a number of line-link and trunk-link networks which are interconnected through the junctor grouping frame. The number of such networks depends upon traffic requirements of the particular electronic central office, and may vary from one to sixteen frames for the line-link and trunk-link networks, respectively. The junctor grouping frame also provides the terminations for the junctor circuits which are used for intraoffice calls through the line-link network.

The major elements of the switching network and the path of an intraoffice call through the eight stages of switching are represented in Fig. 7-7. The path is made up of links connected by remreed switches that were previously described. Each link is essentially a two-wire path consisting of the tip and ring conductors. It is divided into two stages for switching purposes, which are designated 0 and 1 as shown in Fig. 7-7.

In the line-link network, the line switching frames perform the first two stages of switching by connecting the customer lines on their A link inputs to the B links associated with the outputs of the A links. The junctor switching frames, in both the line-link and trunk-link networks, likewise perform two stages of switching. Their C links interconnect the B links with the junctor grouping frame on intraoffice calls in order to provide a path to the trunk switching frame. Two stages of switching also are completed by the A links in the trunk switching frame to connect to the selected outgoing

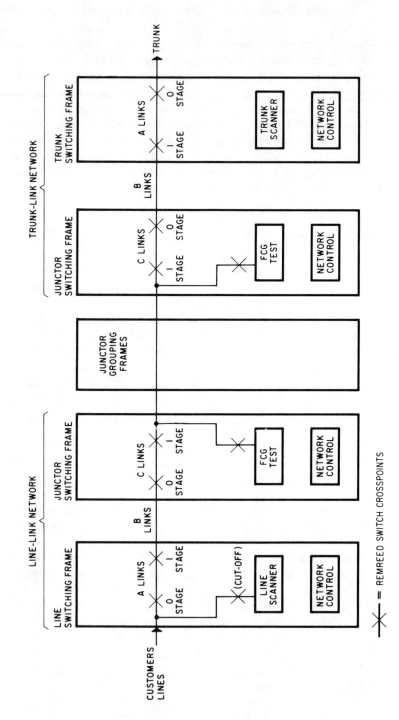

Fig. 7-7. Major Elements of Switching Network in No. 1 ESS

trunk. The FCG test refers to the false cross and ground tests previously explained. These tests are made on a segment of a path before it is closed through the network, to ensure that the conductors are not crossed or grounded.

The line switch frames also accomplish the traffic concentration function of the network. Two different concentrator ratios are usually provided by the line switching frames between the inputs from customer lines and the outputs to the B links. These ratios are 4 to 1 and 2 to 1. The 4 to 1 concentrator line switching frame has sixteen line concentrators. Each concentrator interconnects sixty-four customer lines to sixteen B links through the two switching stages designated 0 and 1 as shown in Fig. 7-7. Thus, one complete line switching frame terminates 1,024 customer lines on its input and 256 B links on its output.

The 2 to 1 line switching frame also has sixteen line concentrators. Each concentrator interconnects thirty-two customer lines to sixteen B links through two switching stages in a similar fashion as in the case of the 4 to 1 concentrator. Therefore, 512 customer lines may be terminated on the inputs of this type of line switching frame for connection to 256 B links. Other concentrator ratios, such as 3 to 1, 5 to 1, 6 to 1, 7 to 1, or 8 to 1 may be obtained by using different combinations of junctor switching frames with the line switching frames.

Figure 7-8 is a simplified block diagram of a line switching frame that is arranged for a 4 to 1 concentrator ratio. Note that a group of sixty-four customer lines is assigned to four input or stage 0 switches, utilizing sixteen lines per switch. Each switch is an array of sixteen crosspoint remreeds. The wiring between stages 0 and 1 is designated the A links. Each of the customer lines on a stage 0 switch can be connected to only four of the eight outputs of the switch. These four outputs of the switch (A links) are wired, one each, to the four switches in stage 1. Figure 7-8 also illustrates examples of wiring between some of these switches. Each line in the drawing represents a tip and ring pair. Access to all B link outputs, consequently, is obtained with a minimum number of crosspoint remreeds per A link.

Note that each customer's line in Fig. 7-8 also connects to a pair of bipolar remreeds or cutoff switches. These cutoff switches provide the connections to the line scanner circuit which initiates requests for service. Control equipment in the line switching frame governs the operation of these bipolar remreed switches.

Trunk and Junctor Switching Frames

The trunk and junctor switching frames are similar in construction except that the trunk switching frame is not equipped with bipolar remreed

Fig. 7-8. Simplified Block Diagram of 4 to 1 Concentrator Line Switching Frame

switches for access to the FCG test circuit. Both types of switching frames perform two stages of switching. The trunk switching frame interconnects trunks and service circuits on its input (A links) with the B links on its outputs. The associated junctor switching frame interconnects the B links with the two stages of its C links. These three interconnecting links or paths are completely within the trunk link network as shown in Figs. 7-7 and 7-8.

Both trunk and junctor switching frames are usually arranged for a 1 to 1 concentrator ratio utilizing four octal grids. Figure 7-9 is a simplified

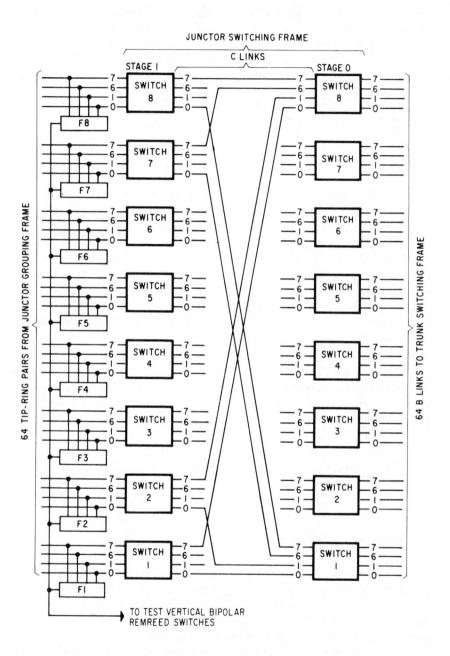

Fig. 7-9. Simplified Block Diagram of Junctor Switching Frame in Trunk Link Network

block diagram of a typical junctor switching frame. This diagram is also applicable to its associated trunk switching frame. The sixty-four inputs or B links are assigned to eight switches in stage 0, and the sixty-four outputs are connected to eight switches in stage 1 by the C links. Each switch handles eight links or tip and ring conductor pairs. The interstage wiring connects each input switch to each of the eight output switches. Therefore, each input or B link will have access to every one of the sixty-four output terminals through a single path defined by a particular C link.

Junctor Grouping Frame

The line-link and trunk-link networks are interconnected through two or more junctor grouping frames. The junctor grouping frame actually provides an interconnecting path between the junctor switching frame of the line-link network and the junctor switching frame of the trunk-link network on all interoffice calls. This path is shown in Fig. 7-7. For intraoffice calls, the junctor grouping frame connects a junctor circuit associated with the junctor frame to the C-link output of the line-link network. Likewise, the junctor grouping frame provides trunk-to-trunk interconnections on calls routed from distant central offices through the particular electronic switching office. The block diagram in Fig. 7-10 illustrates the various interconnections provided by the junctor grouping frame.

The junctors from the line-link and trunk-link networks are arranged in subgroups of sixteen tip and ring conductor pairs. These sixteen pairs are connected via terminal strips on the rear of the junctor grouping frame, either to a connector or to a plug-ended patch cord on the front of the frame. This arrangement is illustrated by the photograph in Fig. 2-7. Thus, any desired interconnection pattern may be made between the link-line and trunk-line networks by the jack and plug system installed in the junctor grouping frame. Equivalent interconnection patterns in electromechanical central offices would necessitate substantial rewiring work.

Establishing Network Paths

Paths through the line-link and trunk-link networks are created by a combination of links and stages of remreed switches operated under the direction of central control. These links and associated wiring form a transmission or talking path when the appropriate crosspoint contacts on the related remreed switches in the network are closed.

Each switching frame in the network contains two controllers which are operated by the central pulse distributor (CPD) under the direction of central control. Every network controller governs the operation of one-half

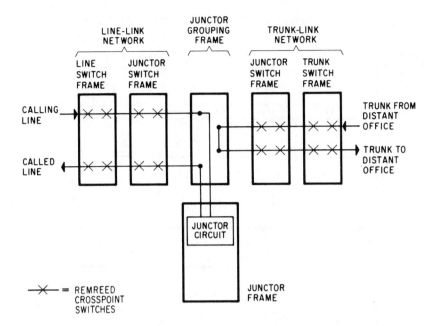

LINE-LINK
NETWORK

JUNCTOR
GROUPING
FRAME

TRUNK-LINK
NETWORK

LINE
SWITCH
FRAME

JUNCTOR
SWITCH
FRAME

JUNCTOR
SWITCH
FRAME

TRUNK
SWITCH
FRAME

CALLING
LINE

TRUNK FROM
DISTANT
OFFICE

TRUNK TO
DISTANT
OFFICE

CALLED
LINE

JUNCTOR
CIRCUIT

JUNCTOR
FRAME

—✕— = REMREED
CROSSPOINT
SWITCHES

Fig. 7-10. Interconnections Provided by Junctor Grouping Frame for Intraoffice Calls and Trunk-to-Trunk Connections

of the remreed switches in the particular switching frame. In normal operation, each network controller functions with its assigned group of remreed switches. Thus, two simultaneous path selections can be made within a switching frame, one under the direction of each controller. If a network controller should fail, its mate will take control of both groups of remreed switches. In this event, only one path selection will occur at a time within that particular switching frame.

To explain the operations involved, let us first consider how paths in the network are produced for an intraoffice call from customer line A to line B as shown in Fig. 7-11. When line A is idle (on-hook) prior to initiating the call, its assigned contacts on the crosspoint remreed in stage 0 of the line switching frame are open. At the same time, the cutoff contacts on the bipolar remreed associated with line A's ferrod sensor will be closed. (Refer to Fig. 2-9 for a schematic diagram of a line's remreed contacts and its ferrod sensor.) Concurrently, the line scanner circuit checks periodically for an off-hook indication to start the necessary actions when customer A originates a call.

As soon as the call to line B has been dialed by calling customer A, central control will initiate the setting up of a path through the line-link network to interconnect the calling and called lines. For this purpose, central

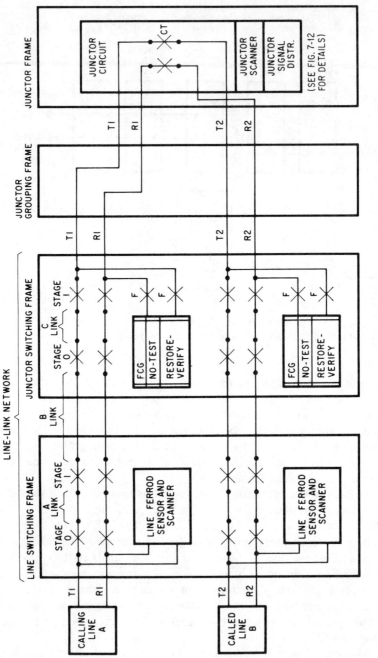

Fig. 7-11. Network Paths Established for an Intraoffice Call in No. 1 and No. 1A ESS

control maintains a current record of the busy-idle states of all network lines on a network map, as per the example shown in Fig. 6-7. Also, central control keeps a record of the end termination of all network paths in use or reserved within the path memory of call store as explained in Chapter 6 and Fig. 6-8.

The network map in call store is examined whenever a path is to be selected between a line and junctor switching frame, or between a trunk and junctor switching frame. The path selection is made by central control on the basis of the busy-idle states of all possible A, B, and C links between the line and a selected junctor group. The same procedure is also followed in choosing a path through the trunk-link network to a selected trunk in the case of an interoffice call.

After the path is selected, the network map in call store is updated and a record of the line junctor (or line junctor trunk association in the case of an interoffice call) is entered in the path memory of call store. Central control next prepares appropriate network instructions. These instructions or work lists, which include the identity of the line link and junctor frames involved, are recorded in a particular peripheral order buffer (POB) in call store. One POB, as explained in Chapter 6 and Fig. 6-6, is associated with each call requiring a connection through the network. The aforementioned instructions subsequently are read out of the POB at 25-millisecond intervals for transmission to the required network controllers.

Transmission Paths for Intraoffice Calls

A partial path first is set up through the line-link network. This is accomplished by closing the remreed switches in stage 1 of the line switching frame, and the remreeds in stage 0 of the associated junctor switching frame. This is indicated in Fig. 7-11 by the T-1, R-1, and T-2, R-2 leads. Next, the F contacts of the bipolar remreeds in the related junctor switching frame are closed to connect these same leads to their respective false cross and ground (FCG) verticals. This permits the connection of the FCG detector circuits which check the conductors (T-1, R-1 and T-2, R-2) in the paths for any foreign potential, false crosses, or grounds.

Assuming that the FCG path tests are satisfactory, central control will direct that the cutoff contacts associated with the line ferrods of both lines A and B in the line switching frame, and their respective F contacts in the junctor switching frame, be opened. Also, at the same instant, the crosspoint contacts of both lines in stage 0 of the line switching frame will be closed. The last step, directed by central control to complete the talking path between lines A and B, is the closure of the cut-through contacts in the assigned junctor circuit. Note that these operational steps apply simultaneously for the paths assigned to both the calling and called lines.

The schematic of a typical junctor circuit on a junctor frame in the No. 1 ESS is illustrated in Fig. 7-12. The junctor circuit furnishes the battery power required for voice transmission and signaling. Separate battery potential is supplied to the calling and called telephone lines for these purposes. Latching relays first were used to close through or open the transmission paths in junctor circuits, as indicated by relays (CT-1) and (CT-2) in Fig. 7-12. They have since been replaced by integrated circuits. These transmission paths are closed after all remreed switches in the line-link network have operated to complete the route between the calling and called lines. (Refer to the paths between lines A and B in Fig. 7-11.)

Thus, no battery potential is present on the remreed crosspoints until after the contacts are closed. On disconnection, the aforementioned latching relays or their equivalent integrated circuits release before the crosspoint contacts of the remreeds are opened. These various operations are controlled by the junctor signal distributing circuit under direction of central control.

The 2 μF capacitors in the transmission path serve to bypass the speech currents. The inductors in series with the T-1, R-1 and T-2, R-2 conductors act as audio-frequency choke coils to keep out the speech currents while allowing the passage of the 48 volt direct current.

Call Supervision by Junctor Circuit

Ferrods 1 and 2 supervise the calling and called line circuits, respectively, for a change of state. The presence or the flow of direct current through the control windings of the ferrods will cause a binary 0 to be read by the junctor scanning circuit, whenever the interrogate winding is momentarily pulsed. The absence of current flow results in a binary 1 reading.

These binary digits advise central control of the status of the call. For instance, a binary 0 from each ferrod indicates that both parties are connected with each other and conversing. Binary 1s are a signal that both parties have hung up. Likewise, a binary 0 from ferrod 1 and a binary 1 from ferrod 2 would advise that only called line B has disconnected. Similarly, a binary 1 from ferrod 1 and a binary 0 from ferrod 2 denote that only calling line A has disconnected.

When calling line A hangs up, ferrod 1 (see Fig. 7-12) sends a binary 1 through the junctor scanning circuit to central control. Consequently, central control will initiate the release of the network paths that had been established for this intraoffice connection, as shown in Fig. 7-11. First, central control causes the opening of the transmission paths in the junctor circuit, by means of the associated integrated circuits. For simplicity, this may be depicted by releasing relays (CT-1) and (CT-2) in Fig. 7-12. The line-link network paths from calling line A to called line B (see Fig. 7-11), however,

NOTE:
THE FUNCTIONS OF (CT1) AND (CT2) RELAYS ARE PERFORMED BY INTEGRATED CIRCUITS
IN LATER NO.1 ESS AND IN THE NO. 1A ESS.

Fig. 7-12. Junctor Circuit Used to Supervise Calling and Called Customer Lines on an Intraoffice Call (NOTE: The functions of (CT-1) and (CT-2) relays are performed by integrated circuits in later No. 1 ESS and in No. 1A ESS) remain closed at this time. Next, central control directs the closure of the bipolar remreeds' cutoff contacts of both lines A and B. These cutoff contacts are in their respective line ferrod sensor and scanner circuits on the line switching frame illustrated in Fig. 7-11. Simultaneously, central control causes the closure of the corresponding F contacts in stage 1 of the junctor switching frame for the connection of the restore–verify circuit.

The restore–verify circuit in Fig. 7-11 will be connected to each of lines A and B by the closure of the F contacts. This circuit connects a resistor across the respective T-1, R-1 and T-2, R-2 leads of the network paths to simulate the origination of a call. This condition will be observed by the line ferrod sensors and scanners of lines A and B, and central control will be informed of the resultant satisfactory path tests. As a result, central control will direct that the crosspoint contacts in stage 0 of the line switching frames be opened to complete the disconnection of the call and to make the network paths available for other calls.

It is possible to monitor conversations at any time during a call by closing the F contacts in the junctor switching frame and connecting the no-test circuit shown in Fig. 7-11. The no-test circuit, however, can be activated only from the master control center. It is utilized for testing purposes and to check for busy-line conditions.

Master Control Center

The operation, maintenance, and administration of electronic central offices that utilize self-diagnostic features are managed by a *master control center (MCC)*. It comprises a number of equipment (hardware) items including a master control console, trunk and line test facilities, automatic message accounting (AMA) recorders, and teletype (teletypewriter) channels. In the No. 1A ESS and later type electronic switching offices, a DATASPEED®-40 terminal with a cathode-ray tube (CRT) display and high-speed printers are provided for improved machine/maintenance personnel communications.

In the original No. 1 ESS, the master control center consisted of four main equipment units as illustrated and designated in Fig. 7-13. A photograph of a typical MCC in a No. 1 ESS office appears in Fig. 2-16. The memory card writer used in the No. 1 ESS, which is described in Chapter 5 and Fig. 5-11, usually was installed adjacent to the MCC. This item is not required in the No. 1A ESS and other ESS types that employ the 1A processor.

Referring to Fig. 7-13, the alarm display and test and control panel of the MCC was the centralized control point of the No. 1 ESS. It contained four sections, each with lamps, keys, pushbuttons, and related rotary switches. The status display lamps showed the system's operating condition.

In case of trouble, craftspersons could quickly ascertain the seriousness of the failure by glancing at lamp indicators. The emergency action section incorporated keys and lamps to take over system control in case of program failure. The communication bus section had keys and lamps for control of the peripheral bus system. In the program interrupt control section, keys could be employed to control the insertion of program data into the system as required during emergencies. Likewise, data from the system could be displayed on lamps in the program interrupt control section.

1 = AUTOMATIC MESSAGE ACCOUNTING RECORDERS
2 = TELETYPEWRITER PANEL
3 = TRUNK AND LINE TEST (TLT) PANEL
4 = ALARM DISPLAY AND TEST PANEL

Fig. 7-13. Major Units of the Master Control Center in No. 1 ESS

In the No. 1A ESS, the master control center normally comprises a master control console frame and an adjacent frame holding a video (CRT) display unit with a standard teletype keyboard (DATASPEED-40 terminal), and an associated high-speed printer. This equipment is utilized for ESS/craftspersons communication and control requirements. A trunk and line test unit, similar to the one used in the No. 1 ESS, also is furnished. It is usually installed in the same aisle with the master control console. The AMA tape recorders are on a separate frame associated with the auxiliary unit bus system. Figure 7-14 is a drawing of a typical master control center arrangement in a No. 1A ESS.

Fig. 7-14. Master Control Center in No. 1A ESS

The primary (program stores and call stores) and secondary (file stores and tape units) memory systems of the 1A processor (see Fig. 2-17) can be controlled from the master control console illustrated in Fig. 7-14. For example, programs can be updated and translation changes can be readily made by craftspersons from the master control console. Also, file stores, which accumulate data on system failures or "interrupts," and on subsequent diagnosis of a faulty unit, are controlled from the MCC. Thus, the file stores can be directed to produce a printout of such failures that occurred during the past 24 hours or longer periods. Likewise, when a trouble analysis reveals certain fault characteristics, the magnetic tape units can be searched to find a known fault with matching traits. Then, a list of related circuit packs may be printed out by the DATASPEED-40 terminal to serve as a guide for maintenance personnel in clearing the trouble.

The master control console (MCC), shown in Fig. 7-14, consists of three panels that are designated: the *process or display panel, processor request panel*, and the *system application panel*. These panels contain keys, lamps, and light-emitting diode (LED) displays as well as terminal connections. The keys, lamps, and other controls on the panels of the MCC are powered and controlled over a cable from the *processor peripheral interface (PPI)* frame. The MCC is associated with the PPI as a subsystem of the 1A processor as depicted in Fig. 2-17.

The *processor display panel*, shown in Fig. 7-15, contains the controls and displays required for controlling the 1A processor used in the No. 1A

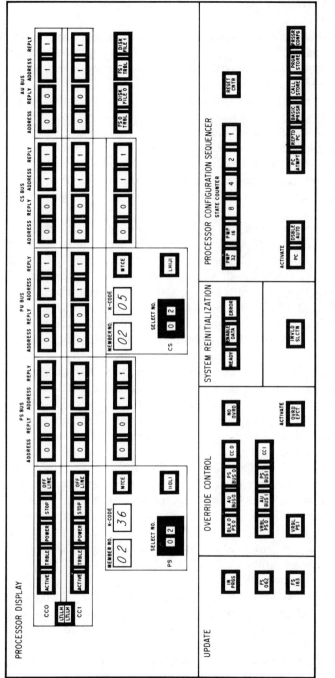

Fig. 7-15. Processor Display Panel of Master Control Console in No. 1A ESS (Courtesy of Bell Telephone Laboratories)

219

ESS, No. 4 ESS, and later type ESS central offices. It is divided into the following five main groups of keys, lamps, and LED displays for this purpose: *processor display, update, override control, system reinitialization,* and *processor configuration sequencer.*

Referring to Fig. 7-15, the update group of keys and lamps permit craftspersons to check when a program update is in progress. The override control set of switches and lamps allow them to manually activate a central control unit, auxiliary bus unit, and program store buses for emergency system recovery. The system reinitialization keys and lamps provide for manual system reinitialization by craft personnel in conjunction with either the override control or processor configuration sequencer group of keys and lamps.

The processor request panel is adjacent to the processor display panel as illustrated in Fig. 7-14. It is divided into the manual interrupt program request and the system display sections. This panel enables maintenance personnel to request displays of various system operating data in binary or numeric form. It also permits craftspersons to monitor or check the status of different indicators and circuits in the master control console. The system application panel, which is located next to the processor request panel, provides information on the condition of the switching network with respect to the requirements of the particular using system.

The trunk and line test or TLT frame in the MCC furnishes the means for testing and removing from service, if necessary, all outgoing trunks, service circuits, and customer lines. Facilities for the disposition of permanent signals, particularly those caused by receivers off-hook, are also included. A master line, associated with a Touch Tone® set, is connected to the line-link network in the same manner as a customer's line. This arrangement is illustrated in Fig. 7-16. The master test line, therefore, can be connected to any trunk or service circuit for test purposes. Consequently, it is not necessary to provide individual test or outgoing test (OGT) jacks for every outgoing trunk as is the case with electromechanical central offices. In addition, any trunk or service circuit can be connected to the voltmeter or transmission test circuits through one of several access trunks as indicated in Fig. 7-16. Any customer line, likewise, can be reached through the line-link network for connection to the voltmeter test circuit. Figure 7-17 shows the network connections to idle and busy customer lines. The no-test circuit, associated with the test vertical in the junctor switching frame (see Fig. 7-11), can only be activated from the master control center to test an off-hook line.

In many telephone operating companies, the maintenance and testing of electronic switching (ESS) and many crossbar-type electromechanical offices are performed by regional *switching control centers.* These centralized operations, which have resulted in substantial cost savings, were made

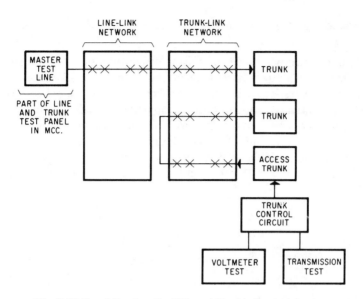

Fig. 7-16. Trunk Testing Facilities of Master Control Center

possible by the use of minicomputers and improvements in electronics technology. For instance, the minicomputer, and associated equipment units in switching control centers automatically monitor remote ESS offices to detect and identify software and hardware faults. In addition, they cause routine tests to be made on various equipment units, analyze operating con-

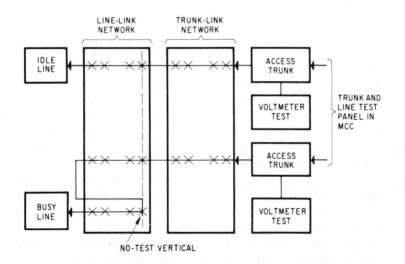

Fig. 7-17. Line Testing Connections from Master Control Center

ditions, and search for malfunctioning components that could deteriorate service.

Maintenance personnel, by means of control consoles in a switching control center, can direct operations in 5 to 10 remote ESS offices just as if they were at the respective master control console in those offices. Teletype data channels are used for communication between craftspersons at switching control centers and remote ESS offices. These teletype channels connect to the input/output frames of the No. 1A processor. A maximum of 96 teletype channels is available for various communication purposes, as per the block diagrams in Fig. 7-18, which indicate some of the major channels.

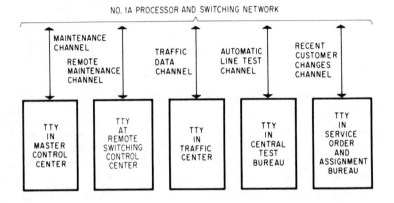

Fig. 7-18. Typical Teletype Channels Used for Communication with Craftspersons in No. 1A ESS

These teletype channels may be equipped with EIA RS-232 asynchronous terminals or the DATASPEED-40 type which was previously described. Data transmissions between these terminals and the master control console in the No. 1A ESS are in the ASCII teletype code (American Standard Code for Information Interchange) as defined by the American Standards Institute. This is an 8-bit code in which each character is represented by a series of 7 binary digits (bits) plus 1 "parity" bit used to detect errors. For asynchronous transmission (the transmission timing depends on the reception of each character); every character is preceded by a "start" bit and followed by one or more "stop" bits for a total of at least 10 bits per character. The number of bits sent in 1 second is termed a *baud*. Therefore, the DATASPEED-40 can transfer data at speeds of 110 or 1200 bauds.

The ASCII method, however, cannot be employed for transferring data between central control and its associated processing equipment, such as program and call stores, which use data in pure binary form. Therefore, two input/output channels can be arranged to transfer data in 8-bit groups. This arrangement would permit transferring data in pure binary form instead of the ASCII teletype system.

Teletype messages from either the local master control center or the switching control center normally are either action or information types. For instance, the system may be asked to trace a call through the network from a specific customer line or trunk. The system may also be ordered to perform a diagnostic test on certain parts. Information messages, for example, may pertain to a list of busy trunks, unusual operating conditions or traffic and administrative data requiring no immediate work. Action messages, on the other hand, require immediate or subsequent operations by maintenance personnel depending on the action category. For example, a major alarm due to a trouble condition in central control requires immediate action. A minor alarm caused by a failure in a service circuit, however, may be deferred for later diagnosis.

The automatic message accounting (AMA) recorder stores on magnetic tapes the charging information pertaining to calling customer lines. These tapes are subsequently transported to a data processing center for use in computing customer bills. One 2,400-foot reel of tape is usually adequate for storing the billing data for about 100,000 calls in one day. A duplicate AMA unit is provided in the master control center to ensure reliability and recording continuity. The sequence followed for recording AMA data on a call is illustrated in Fig. 7-19. If the call processing program determines that AMA data is to be recorded on a particular call, the pertinent information will first be temporarily stored in an AMA register in call store. Upon completion of the call, the related data are assembled in the binary-coded decimal (BCD) format, as explained in Chapter 3 and Table 3-1, and

Fig. 7-19. AMA Data Flow Diagram

then placed in an AMA buffer in call store. The recording procedure is initiated by an AMA program in program store when the AMA buffer in call store is fully loaded. This AMA program will cause central control to direct that the data be transferred, one word at a time, to the AMA circuit for recording on the magnetic tape. This procedure is shown by the block diagrams in Fig. 7-19.

Power Plant System

The power plant system for the No. 1A ESS and similar electronic central offices includes two common battery power supplies. One supplies – 48 volts and the other furnishes + 24 volts to the power distributing frames in the central office. Both battery plants can handle voltage tolerances of plus or minus 10 percent. This design avoids the need for end cells and counter cells that are provided in battery plants of electromechanical central offices in order to maintain stable voltage conditions. The normal input power source is commercial ac power at 208/120 volts, 60 Hz, 3-phase, which employs four-wire transmission. To protect against possible prolonged failures of the commercial ac power, one or more engine-alternator sets are installed.

A simplified block diagram of a typical power system for electronic switching offices appears in Fig. 7-20. The + 24 volt and the − 48 volt storage batteries have sufficient capacities to handle traffic loads during busy hours in case of an ac power failure. The − 48 volt battery, likewise, supplies power to the coin control and the ringing and tone supply power plants. It also has adequate capacity for operating the emergency ac power plant which provides 3-phase, 208/120 volts, 60 Hz power to the ringing plant, announcement machines, teletypewriters, and the AMA magnetic tape recorders in the event of a commercial ac power interruption. This emergency ac power plant starts automatically whenever an ac power failure occurs.

Three power feeders, − 48 volts, ground, and + 24 volts, as shown in Fig. 7-20, are run from the power room to power distributing frames in the switch room. Each battery feeder is filtered by a 35,000 μF capacitor bank. Distribution of power to the equipment frames is arranged so that no two duplicate circuits are connected to the same power distributing frame. Moreover, to minimize electrical noise, all frames and cable racks are insulated from the building at the time of installation. Power plant ground returns are similarly insulated from their power plant frames.

The 20-Hz ringing voltage is usually produced by a duplicate set of ringing generators. The audible ringing tone and all other tones are derived from four fundamental tones, generated by precision transistor oscillators, and amplified. These fundamental tone signals are 350 Hz, 440 Hz, 480 Hz,

Fig. 7-20. Power Systems Including dc Power Supplies for Electronic Central Office

and 620 Hz. All utilized tones, except the high tone, consist of a combination of two fundamental tones. The actual output of each transistor oscillator is a square wave. However, it is converted by bandpass filters to a sine wave, with a harmonic level at least 60 dB down from the basic frequency, in order to produce the required precision tone. Table 7-1 lists the various tone designations and their precise frequencies as employed in electronic central offices.

Table 7-1. Designated Tones and Frequencies Used in Electronic Switching Offices

Tone Designations	Frequencies
*Dial tone	350 Hz plus 440 Hz
*Audible ringing tone	440 Hz plus 480 Hz
*High tone	480 Hz
*Low tone	480 Hz plus 620 Hz
Line busy tone	Low tone at 60 IPM
Path busy tone	Low tone at 120 IPM
Ringing current	20 Hz

*Fundamental tones
IPM = Interruptions per minute

Questions

1. Why is it necessary to employ switching devices that physically open unused parallel paths in electronic central offices?
2. How many switching stages are provided in the No. 1A ESS switching network? Where are they located?
3. What are the main components of the crosspoint remreed switch? Where is it employed?
4. How is the remreed switch similar to the ferreed? What are the main advantages of the remreed switch?
5. What are the major elements of the switching network and what frames comprise each element? What traffic concentration ratios are normally provided and where?
6. What links and frames are in the transmission path from the calling to the called line on an intraoffice call? On an interoffice call?
7. What binary digits read from the ferrods in the junctor circuit indicate that both parties are connected with each other? What binary digits indicate that both parties have disconnected? Why?
8. What are the major elements of the master control center in the No. 1A ESS? What panels comprise the master control console? What central means are used to reduce maintenance costs of ESS offices?
9. What kinds of teletype messages are normally sent over maintenance channels? Give an example of each kind of message. What type of terminal equipment is normally employed for these teletype channels?
10. Which batteries make up the dc power system in an electronic central office? What are the fundamental tone signals from which all other tones are derived? What frequencies are used for dial tone and for the line busy tone?

8

Electronic Switching Systems for Suburban and Rural Areas

Growth of Electronic Switching Systems

The basic concepts and the hardware and software fundamentals of electronic switching systems are exemplified in the foregoing descriptions of the initial No. 1 ESS and the improved No. 1A ESS. In essence, these end central offices (Class 5 in Fig. 1-2) incorporate high-speed data processors, large-capacity semipermanent and temporary memory systems, 2-wire magnetic reed switches, and stored program control operations. The data processing is directed by central control, which functions with the semipermanent (program store) and temporary (call store) memories. The program instructions and translation information are in binary digital formats.

This coalition directs the call-processing operations of the above-mentioned electronic switching systems on a specified time-shared basis. Moreover, stored program control provides swift and economical means for making additions and changes in subscriber lines, class of service, trunks, and related call-handling functions. Such modifications usually are completed without hardware or wiring alterations.

Electronic telephone switching systems employ computer-type technologies. The stored program or memory phase of a general purpose computer and the program store in the No. 1A ESS are similar, but their dependability requirements differ considerably. For instance, a temporary stoppage of the computer may delay the output results which would usually be an inconvenience only to the user. Any stoppage of the electronic switching system, however, would halt telephone service in the central office which would be a disastrous situation. On the other hand, a data processing error in the No. 1A ESS may result only in a wrong number on the particular call being processed. In a computer, a processing error cannot be tolerated because it may cause, among other things, an overcharge in an account. It is seen, therefore, that the dependability requirements of electronic switching systems are very high. This has necessitated the duplication of many subsystems and peripheral units. Such equipment duplication is not usually required in computers.

The Bell System's No. 1 ESS and No. 1A ESS, previously described in this book, are capable of serving many tens of thousands of lines in large cities and metropolitan areas. For instance, the No. 1 ESS can handle approximately 110,000 calls in a peak busy hour, and the No. 1A ESS handles about 240,000 peak busy-hour calls. Consequently, it would not be economical to utilize these large-capacity switching systems for the telephone requirements of small localities in suburban and rural areas. Therefore, abridged and more economical versions were evolved by the Bell Telephone Laboratories for these particular applications. They are designated No. 2 ESS and No. 2B ESS for suburban areas, and the No. 3 ESS for rural communities. Descriptions of these electronic switching systems are presented in this chapter.

Furthermore, the basic design of the Bell System's No. 1 ESS and the improvements in the No. 1A ESS led other laboratories and equipment manufacturers to produce electronic switching equipment predicated on similar techniques. These techniques included the use of high-speed electronic central processor operating with stored program control of switching operations on a time-shared basis. Table 8-1 lists some of these other major electronic switching systems having the above characteristics and also the current Bell System types. It is not within the scope of this book to attempt to explain all of the systems indicated in Table 8-1. Hence, only succinct de-

Table 8-1. Major U.S. Electronic Switching Systems

System Designation	Manufacturer	Maximum Terminals	Maximum Busy-hour Calls
No. 1 ESS	Western Electric Co.	65,000	110,000
No. 1A ESS	Western Electric Co.	128,000	240,000
No. 2 ESS	Western Electric Co.	19,000	19,000
No. 2B ESS	Western Electric Co.	20,000	38,000
No. 3 ESS	Western Electric Co.	4,500	11,000
No. 4 ESS*	Western Electric Co.	107,000	550,000
No. 1 EAX	GTE Automatic Electric Inc.	64,000	79,000
No. 2 EAX	GTE Automatic Electric Inc.	40,000	60,000
No. 3 EAX*	GTE Automatic Electric Inc.	60,000	350,000
NX-1E†	North Electric Co.	20,000	68,000
TCS-5‡	ITT-Telecomm.	6,000	23,000

*Digital switching system for tandem and toll (Class 4) offices.
†Utilizes crossbar switches for switching network.
‡Switching is accomplished by diode matrixes.

scriptions are presented of the Bell System's No. 2, No. 2B, and No. 3 ESS offices, and the No. 1 and No. 2 EAX of GTE Automatic Electric Inc.

It should be noted, with respect to the above-mentioned switching systems, that speech or voice-frequency currents are generated in the continuous wave or analog mode. Electronic switching systems, therefore, were first designed for analog switching utilizing the space-division method. That is, each call is routed over separate selected metallic paths through the switching network. Thus, spacious hardware items and numerous control circuits are required at sizable costs for analog switches. Recent developments in semiconductor technology, particularly large scale integrated (LSI) circuits, have made it feasible to employ digital switching at substantial cost savings. Thus, the No. 4 ESS (listed in Table 8-1) is the first digital switching system developed by the Bell Telephone Laboratories for toll (Class 4) centers. The No. 4 ESS and other digital telephone operations are discussed further in Chapter 9.

No. 2 ESS and No. 2B ESS Design Precepts

The Bell System's No. 2 ESS was developed shortly after the introduction of the original No. 1 ESS. The No. 2 ESS was designed primarily for use in nonmetropolitan areas to serve approximately 1,000 to 19,000 customer lines. It incorporated the major precepts of the initial No. 1 ESS and was capable of unattended operations. Subsequent advances in semiconductor technology and the evolution of the 1A processor (used in the No. 1A ESS as explained in Chapter 5) produced the improved No. 2B ESS. The No. 2B ESS requires much less space and power consumption than the No. 2 ESS. It can handle 38,000 peak busy-hour calls, which is twice as many as the No. 2 ESS.

The existing switching network, line scanners, and trunk circuits of the No. 2 ESS are used by the No. 2B ESS. It also employs an augmented 3A processor from the No. 3 ESS, which is designated the 2B processor. The 2B processor contains integrated circuits instead of the individual transistors, diodes, and other components that are utilized in the processor of the No. 2 ESS. Consequently, the No. 2B ESS has double the call-processing capacity because its 2B processor operates more than twice as fast as that of the No. 2 ESS. Therefore, information can be read from the stored memories faster and the logic circuits in central control can interpret the data more rapidly. For brevity, the descriptions in this chapter will apply to both the No. 2 ESS and the No. 2B ESS unless otherwise indicated.

The service features offered by both No. 2 and 2B ESS offices are essentially the same as those described in Chapter 2 for the No. 1 ESS. However, the No. 2 and 2B ESS also furnish four-party and eight-party line service, which still is needed in some suburban and rural areas. Both the No. 2 and 2B ESS have a maximum capacity of 19,000 line terminals. The designated traffic capacity of the No. 2 ESS is 19,000 peak busy-hour calls, while the No. 2B ESS can handle up to 38,000 calls because of its higher call-processing speed and related factors.

Many of the frames and apparatus developed for the No. 1 ESS are also utilized in the No. 2 and No. 2B ESS. For instance, the peripheral equipment such as remreed switches, ferrod scanners, and the trunks and service circuits of the No. 1 ESS are included in the No. 2 ESS as well as the No. 2B ESS. Likewise, the program and call stores in the No. 2 and 2B ESS employ the same memory devices and operate in a manner equivalent to their counterparts in the No. 1 and 1A ESS, respectively. Consequently, only the apparatus and functions of the No. 2 ESS not previously explained will be specifically described in this chapter. (The reader is referred to other chapters for applicable details concerning equipments and functions common to the No. 1 and 1A ESS and the No. 2 and 2B ESS central offices.)

Basic Elements of No. 2 and No. 2B ESS

The major components of the No. 2 ESS consist of the switching network, master scanner, two central processor frames, and the maintenance control center as portrayed in Fig. 8-1. The No. 2B ESS has similar major components except that the *2B processor* replaces the central processor frames as subsequently explained.

The number of frames and equipment quantities in both No. 2 and 2B ESS are greatly reduced in comparison with the respective No. 1 ESS and No. 1A ESS. For example, the major equipment frames are the line-trunk switching, network control junctor switching, junctor grouping, central processor, and the maintenance control center with its display and trunk test panels, and teletype terminals. Other important peripheral equipment frames are the master scanner, supplementary central pulse distributor, automatic message accounting (AMA), and various frames containing trunks, service, and junctor circuits. Most of these major equipment components and their correlations are shown in the block diagrams of Fig. 8-2.

The two program and processing control units of the No. 2 ESS, illustrated in Fig. 8-2, synchronize with each other. Control unit 1 normally is "on-line" or in command of the call processing functions. The operations of both control entities are matched at the maintenance control center. Thus, in case of any failure, control unit 2 instantaneously takes over operations. The call store in each control unit is shared by both the program and processing control and the input-output control circuit. Each control unit

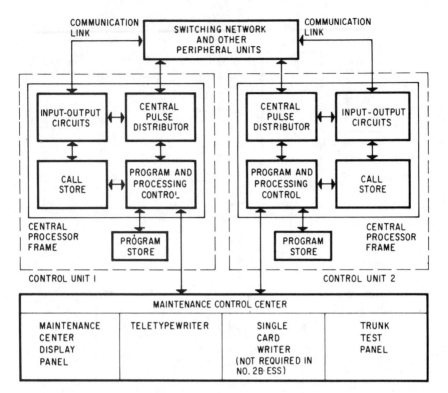

Fig. 8-1. Control Entities of No. 2 ESS

connects to the switching network and other peripheral units over a peripheral bus system or communication link. These buses, as later explained in Fig. 8-8, provide access to the duplicate controllers of the various peripheral frames, such as the master scanner, line-trunk switching, network control junctor switching, and the trunk frames.

No. 2 ESS Central Processor Functions

The central processor frame in the original No. 2 ESS, a photograph of which appears in Fig. 8-3, performs the major call processing functions. Its main components, as depicted in Fig. 8-2, are the program and processing control unit, call store, input-output control circuits, and the central pulse distributor. The program and processing control unit directs the processing of calls in a manner equivalent to central control in the No. 1 and 1A ESS. It also detects and diagnoses equipment malfunctions with the aid of the maintenance center.

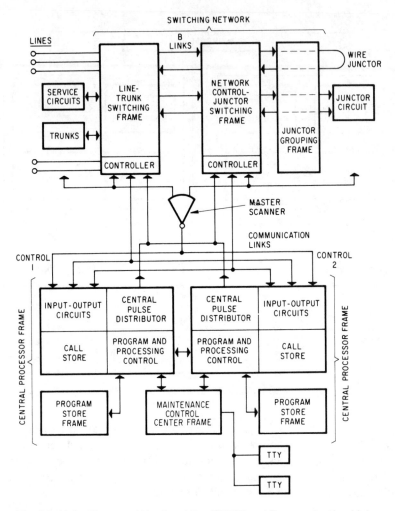

Fig. 8-2. Major Frames of No. 2 and No. 2B ESS and Communication Links

The central processor in the No. 2 ESS contains logic circuits to write and read up to 32,768 binary words (each of 16 bits) of information that can be temporarily stored in up to four call stores. Two of these call store units are installed in the central processor frame, and the other two, when required, are mounted in a supplementary call store frame. The call store employs the same one-inch square ferrite sheet memory device originally used in the No. 1 ESS call store, which is described in Chapter 5. In the No. 2B ESS, one call store is included in each of the main stores of the 2B processor, as shown in Fig. 8-8.

Fig. 8-3. Central Processor of No. 2 ESS (Courtesy of Bell Telephone Laboratories)

Call store operations are similar in many respects to those in the No. 1 ESS as described in Chapter 5, particularly Figs. 5-14 through 5-16. Temporary storage for a random-access memory of 8,196 binary words of 16 bits each is provided in a call store. There are no registers in call store because direct wire connections are used to receive address and input data from pertinent registers in the program and processing control circuit. Likewise, output data from call store is sent by dc pulses directly to registers in either the input-output circuit or to the program and processing control circuit, depending upon the readout instructions. These register relationships and other elements of call store are depicted in Fig. 8-4.

The program and processing control and other central processor units employ large numbers of binary logic circuits in a similar fashion to central control in the No. 1 ESS. The basic logic circuit used in the initial No. 2 ESS, however, was the transistor-resistor NOR gate. A schematic drawing of this logic gate with approximate resistor values is illustrated in Fig. 8-5.

*CALL STORE IS PART OF MAIN STORE IN 2B PROCESSOR.

Fig. 8-4. Call Store Operation Registers in No. 2 ESS

Fig. 8-5. Basic Transistor-Resistor NOR Logic Gates in No. 2 ESS

The gate has two inputs and a maximum of three outputs or fanouts. NOR gate operating principles are explained in Chapter 3 and illustrated in Figs. 3-8 and 3-9. In the original No. 2 ESS, the NOR logic and similar gates were constructed with high-speed silicon transistors and special thin film resistors. Many variations of the basic NOR logic gate were developed for the different operational requirements of the central processor. Logic gates may have several inputs and as many as ten or more fanouts. A schematic of a typical high fanout transistor-resistor NOR gate is shown in Fig. 8-6. Logic gates also form the binary counter-shift registers that are employed in units of the central processor frame.

The input-output control circuits perform the interface functions between the program and call processing control unit and the peripheral units. Moreover, the input-output control circuits act as part of the program and call processing control unit by providing the necessary facilities for communicating with peripheral units under control of relevant program instructions. These circuits also provide wired logic for independently performing parts of certain tasks, such as digit receiving, digit and data transmission, and line scanning. To conserve program time, wired logic is used for many

Fig. 8-6. High Fanout Transistor-Resistor NOR Logic Gate in No. 2 ESS

simple and highly repetitive tasks, instead of using instructions stored in program store.

The central pulse distributor (CPD) located in the central processor frame (see Figs. 8-1, 8-2, and 8-7) provides the means of steering control pulses to one of 512 outputs of the peripheral units. The central pulse distributor has electrical characteristics similar to the ones in the No. 1 and 1A ESS. Its output pulses are used to select and enable a peripheral unit, such as a line-trunk switching frame, or a trunk or junctor circuit, to receive information from the address bus, and to transmit data back over the scan answer bus to the input-output control circuit. Figure 8-7 depicts the various peripheral bus functions. Balanced twisted pairs, which are transformer-coupled to and from the associated circuits, are used for the peripheral bus system. Operations can be compared to the equivalent buses of the No. 1 and 1A ESS as explained in Chapter 4 and shown in Fig. 4-13.

2B Processor in No. 2B ESS

The *2B processor* furnished in the No. 2B ESS uses large-scale integrated (LSI) circuits, and its memories contain insulated-gate field-effect transistor (IGFET) integrated circuits. Consequently, the 2B processor is considerably smaller than the central processor employed in the No. 2 ESS. For instance, only four equipment bays are needed for the 2B processor as compared to 21 bays for the processor in the No. 2 ESS. This small size also makes it feasible to install the 2B processor in No. 2 ESS offices as a means of upgrading its service to that of the No. 2B ESS.

The major equipment elements of the 2B processor are illustrated in Fig. 8-8. The logic circuitry of its two central control units provide self-checking functions. Thus, the central control units need not operate in synchronism as in the case of the No. 2 ESS. One central control unit stays in a standby (off-line) state, while the other is in operation (on-line). Therefore, if the operating unit should fail and its switched off-line, the standby unit immediately goes on-line to ensure uninterrupted service.

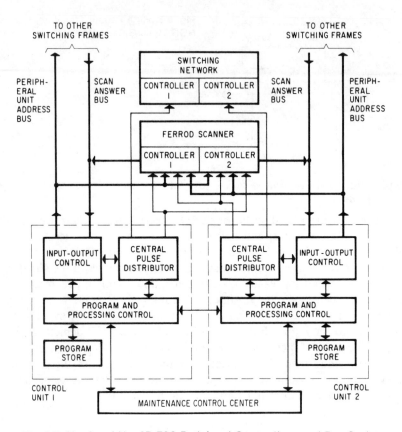

Fig. 8-7. No. 2 and No. 2B ESS Peripheral Connections and Bus System

The design of the 2B processor is based on the technology used in the 1A processor of the No. 1A and No. 4 ESS. Furthermore, the central control units of the 2B processor are adapted, with necessary modifications, from the 3A processor in the new No. 3 ESS. Each central control includes a read-only memory (ROM), logic circuits with 16-bit registers, and interface circuits with main store. Permanent read-only sets of instructions, termed *microprograms*, are stored in the ROM of central control to decode system operating instructions from main store. Likewise, a *microprogram* to direct startup operations, timing and emergency procedures is permanently stored in the ROM of central control.

The two main stores (one is for backup) contain the system's generic program, office data on customer lines, and translation information. Each main store also contains a call store. The call store has a transient memory (electronic slate per Fig. 2-4) for temporarily storing data during the progress of a call. It functions in a similar fashion as call stores in the No. 1 and No. 1A ESS, which is explained in Chapter 5.

The IGFET memory cells of main store initially had a design capacity of 262,166 binary words, each of 16 bits. In order to speed up operations, as required for No. 2B ESS offices, main stores have been modified to store binary words of 24 bits instead of 16. This increase to 24-bit binary words also allows the continued use of existing programs that were written for the No. 2 ESS. Moreover, more operations can be performed in the same time interval. Therefore, it is possible for the No. 2B ESS to control two calls for every one handled by the No. 2 ESS. Furthermore, main store of the improved 2B processor has a larger temporary memory in its call store. This is necessary in order to conform to the higher call-handling needs of the No. 2B ESS.

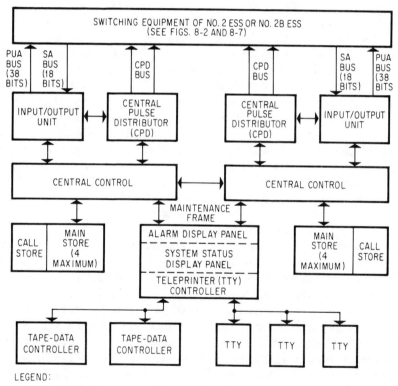

LEGEND:
CPU = CENTRAL PULSE DISTRIBUTOR BUS
PUA = PERIPHERAL-UNIT ADDRESS BUS
SA = SCAN-ANSWER BUS

Fig. 8-8. 2B Processor Equipment and Bus System Connections in No. 2B ESS

In the event of a total power failure, the IGFET memory cells in main store will lose their permanent data. To protect against this contingency, the same data is stored in duplicate magnetic tape cartridges. These magnetic tapes and their transports are located on the *maintenance frame*. Thus, data can be quickly transferred from magnetic tape to the memory cells in main store in such an eventuality. These operations are akin to those of the tape frames in the auxiliary data system of the 1A processor, which was described in Chapter 5.

Another improvement in the 2B processor was the addition of a new *input/output unit*. This unit acts as a buffer between the high-speed central control operations and the relatively low-speed switching equipment in the No. 2B ESS. The input/output unit receives control data from central control and transmits it to the switching equipment over a 38-bit wide *peripheral-unit address (PUA)* bus. Likewise, data from the switching equipment is accepted by the input/output unit over the 18-bit wide *scan-answer (SA)* bus, and then forwarded to central control.

The data transmissions to and from the input/output unit may be represented in an analogous manner. Let us assume that each binary digit of the data is an automobile. Thus, there would be 38 cars on the PUA bus and 18 cars on the SA bus, all traveling abreast on their respective wide highways. In contrast, on the one-lane road between central control and the input/output unit, the cars would be traveling behind each other.

The other interface between central control and the switching network is the 512-point *central pulse distributor (CPD)* illustrated in Fig. 8-8. This equipment unit connects control signals from central control to the controllers and other devices in the switching frames (see Figs. 8-2 and 8-7). These control signals serve to operate trunk, junctor, and service circuits, and to close and open network paths for completing or releasing calls as ordered by central control. Likewise, the various ferrod sensors in these circuits send data concerning the on-hook and off-hook state of customer lines, trunks, junctor, and service circuits to central control via the CPD. These scanning and CPD functions are analogous, respectively, to those described in Chapter 2 (see Fig. 2-2) and Chapter 4 (see Fig. 4-13) for the No. 1 ESS.

The *maintenance frame* is an integral part of the 2B processor. Its appearance is similar to the maintenance control center of the No. 2 ESS that is illustrated in Fig. 8-16. However, it is not equipped with a card writer as portrayed on the left side of the frame in Fig. 8-16. The major components and interfaces of the maintenance frame are shown in Fig. 8-8. The *tape-data controller* comprises the magnetic tape transport and control circuits that govern these functions. The alarm display and system status display panels indicate the performance of major units of the No. 2B ESS. Communications between craftspersons and the No. 2B ESS are conducted, in similar fashion as in the case of the No. 1 and No. 1A ESS, by means of teletype terminals. The teletype equipment and related CRT (video) displays and printers (DATASPEED-40 or similar types) are associated with the maintenance frame.

The No. 2B ESS normally operates unattended, but it is constantly monitored and controlled from a distant switching control center. These operations are conducted over teletype channels in a similar manner as indicated in Fig. 7-18 for the No. 1 and No. 1A ESS.

Program Store Operation in No. 2 ESS

The semipermanent information and memory data for control units 1 and 2 in the No. 2 ESS are furnished by their respective program stores. Program store memory consists of the random-access type but is designed for read only.

The memory device initially employed in program store of the No. 2 ESS was the same type originally provided in the No. 1 ESS. A photograph of this memory mechanism is shown in Fig. 8-9. Its operations are explained in Chapter 5, particularly in Figs. 5-4, 5-6, 5-7, and 5-9. Information in this

Fig. 8-9. Program Store Unit of No. 2 ESS (Courtesy of Bell Telephone Laboratories)

program store could be changed by withdrawing the aluminum cards from the twistor module, and then changing the state of the particular miniature magnets just as in the case of program stores in the original No. 1 ESS. Semiconductor memories, using large-scale integrated (LSI) circuits, have replaced the magnetic types in the later No. 2B ESS as was described.

A program store in the No. 2 ESS contains four memory units and various associated equipment control circuits as shown in Fig. 8-10. Each unit in the original processor of No. 2 ESS included a twistor wire memory module and its related address decoder and access circuits. The program store connects to the program and processing control unit in the central processor frame over the address and answer bus systems. Address data for locating the particular memory unit, and the desired program instructions or translator information in it, is received over the address bus by the 18-bit address register. The information read is first amplified and then transmitted back to the control unit over the answer bus.

To detect circuit malfunctions quickly, program store has checking circuits which verify correct operations on every read cycle. This verification is accomplished by checking the access circuit's current flow in the memory module, and seeing that the correct current path has been selected.

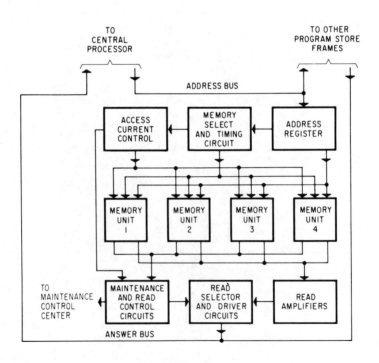

Fig. 8-10. Program Store Operations in No. 2 ESS

In case of a failure, an error signal is sent to the program and processing control circuit and to the maintenance center. Also, as in call store, dc control signals are sent by the maintenance center to program store to simulate different types of malfunctions that the check circuits are designed to detect. In this way, the integrity of the check circuits is verified.

The basic 22-bit instruction words in program store memory consist of two types. One type contains a 5-bit operation code, a 16-bit address, and one transfer check bit. There are relatively few instructions of this first type of instruction word and they are used mainly for program transfers. The other type, which contains most of the instructions, consists of two 10-bit instructions, each with a 5-bit operation code and a 5-bit address, one transfer check bit, and a parity check bit. This instruction format is shown in Fig. 8-11.

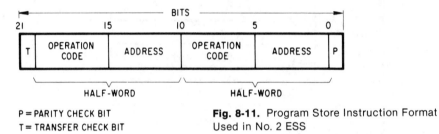

P = PARITY CHECK BIT
T = TRANSFER CHECK BIT

Fig. 8-11. Program Store Instruction Format Used in No. 2 ESS

The central processor can decode and execute only one instruction at a time, just as central control does in the No. 1 and No. 1A ESS. Consequently, the 22-bit instruction format, read from program store, is first temporarily stored in an output buffer until the preceding instruction has been completed. The current instruction is then gated to the program store output register for decoding and completion. Next, the program address register is advanced by 1 so that the following sequential binary word will be read out from program store. In this manner, the succeeding program word can be read while the current instruction in the program store output register is being executed, thereby reducing operating time.

Instructions placed in the program store output register are translated by a two-stage decoding arrangement. This consists of the main decoder and one of eight auxiliary decoders as illustrated in Fig. 8-12. The main decoder selects a particular auxiliary decoder. The selected or enabled auxiliary decoder then translates the 5-bit address to one out of thirty-two connections that lead to the desired program instruction. Outputs of the main decoder also connect to the command and timing logic circuits in order to control the transmission of instructions from program store to the decoders, and to advance the program address register by 1 as previously explained. Subsequent program store operations are similar in many respects

Fig. 8-12. No. 2 ESS Program Store Decoding Arrangement

to the program store functions of the No. 1 and No. 1A ESS as described in Chapter 6.

The No. 2 and No. 2B ESS Switching Network

The No. 2 ESS switching network (also used in No. 2B ESS), shown in Fig. 8-2, is of the space division type similar to the No. 1 and No. 1A ESS. It consists of from one to four line-trunk switching frames (see Fig. 8-13), each associated with a network control junctor switching frame (see Fig. 8-14). The normal capacity of the No. 2 ESS is fifteen line-trunk networks, which can provide up to 30,720 terminals, assuming a 4 to 1 concentration ratio. Eight stages of remreed switches, which form a four-stage folded network, are used for interconnecting the two-wire (tip and ring) metallic paths in the switching network.

Unlike the No. 1 and No. 1A ESS switching network, customer lines, trunks, and service circuits in the No. 2 ESS are assigned to terminals on the same side of line-trunk switching frames. These terminals may be intercon-

Fig. 8-13. Line-Trunk Switching Frame in No. 2 and No. 2B ESS
Courtesy of Bell Telephone Laboratories)

nected by means of the aforementioned eight switching stages and through wire junctors or junctor circuits, depending upon the type of call. The first two and the last two switching stages are situated on the line-trunk switching frame. The third, fourth, fifth, and sixth switching stages are located on the network control junctor switching frame. This eight-stage switching arrangement is illustrated in Fig. 8-15.

The line-trunk network, under control of the central processor, provides the necessary paths to establish intraoffice and interoffice calls, as well as connections to service circuits and to the maintenance center. For instance, an interoffice call is shown in Fig. 8-15 by the paths in the line-trunk network, from calling line A through switching stages 1–4, the wire junctor, and back through switching stages 5–8 to outgoing trunk B. Each line in this drawing represents a tip and ring metallic pair. A line ferrod and scanner circuit is connected to line A through the cutoff contacts, designated C, of its associated remreed switch.

Fig. 8-14. Network Control Junctor Switching Frame in No. 2 and No. 2B ESS (Courtesy of Bell Telephone Laboratories)

A remreed switch, designated F on the network control junctor switching frame, is used to connect a B link to the test vertical on this frame. This test vertical may be connected, under control of the central processor, to either the false cross or ground (FCG) test circuit or to the no-test circuit. The FCG test circuit checks the assigned network paths for false crosses, grounds, or any foreign potentials. The no-test circuit enables maintenance personnel to check for false busy conditions on lines and trunks. These operations are analogous to those performed in the No. 1 and No. 1A ESS as explained in Chapter 7 and Fig. 7-9.

An example of an intraoffice call is also depicted in Fig. 8-15. It is represented by the paths from calling line C through the four switching stages, numbered 1 to 4, in the line-trunk network to the junctor circuit, and from the junctor circuit back through switching stages, numbered 5 to 8, to called line D. A junctor circuit is employed on all intraoffice calls in order to pro-

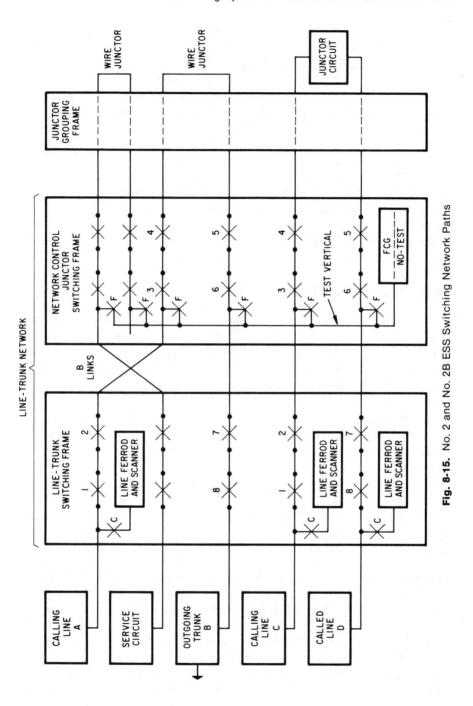

Fig. 8-15. No. 2 and No. 2B ESS Switching Network Paths

vide talking battery and supervision on these calls. Trunk circuits provide these facilities for interoffice calls so that only wire junctors (metallic connections) need be utilized for such calls.

The switching network is controlled by a pair of controllers in a similar fashion as in the No. 1 and No. 1A ESS. Each controller has access to the entire network. Both network controllers may operate simultaneously, provided that the two paths being connected are in different concentrator groups and not in the same switching stages. Network controllers receive path information and other data from the central processor. The data is transmitted, in the form of bipolar binary pulses, by the central pulse distributor as indicated in Fig. 8-7. Internal check circuits monitor operations of the network controllers to preclude malfunctions. Three scan points are associated with each controller for this purpose. These scanner points, which are terminated at the master scanner shown in Fig. 8-2, indicate the state of the network controller at any time.

Maintenance Control Center in No. 2 ESS

The major elements of the maintenance control center in the No. 2 ESS are shown in Fig. 8-16 as follows: the display panel, trunk test panel, teletypewriter, and the single card writer. The separate trunk test frame is not shown in this photograph. Interconnections of the maintenance control center with control units of the No. 2 ESS are portrayed in the block diagrams of Figs. 8-1 and 8-2.

The maintenance center frame contains the equipment for the manual control of the No. 2 ESS. It also includes display panel controls, teletypewriter, and associated interface units for communicating with maintenance personnel. The maintenance center serves primarily as a system maintenance tool to ensure continuous and efficient performance of the No. 2 ESS. In this respect, its principal functions are to monitor the operational status of the system, to provide for manual control of operations in the event of failure of diagnostic procedures or in emergency situations, and to provide for routine test and related control operations.

In addition to the teletypewriter furnished in the maintenance center frame, up to a total of eight teletypewriter channels can be handled by the No. 2 ESS. Each such channel includes a tip and ring loop control circuit and connections to a data set. Teletypewriter channels may be established to a remote switching control center, test bureau, traffic center, and the assignment bureau, just as explained in Chapter 7 and Fig. 7-18 for the No. 1 and 1A ESS.

The trunk test frame, usually installed adjacent to the maintenance center frame, is used to make operational and transmission tests on trunks

Fig. 8-16. Maintenance Control Center of No. 2 ESS (Courtesy of Bell Telephone Laboratories)

tests on trunks and customer lines. The test equipment for this purpose includes transmission and noise measuring sets, impulse counter, peak-average ratio meter receiver, and a voice-frequency oscillator. Complete functional testing of customer lines is usually accomplished by separate test facilities, such as the No. 3 local test unit.

The maintenance center frame may also be equipped with a single card writer, as shown on the left side of the frame in Fig. 8-16, for only No. 2 ESS offices. The card writer consists of a writing head and means for moving the head over an aluminum memory card. The card writer functions in a manner similar to the card writing equipment installed in the No. 1 ESS and described in Chapter 5. It is not required for No. 2B ESS installations.

No. 2 and No. 2B ESS Power Plant

The design of the power plant in the No. 2 and No. 2B ESS is very much the same as that described for the No. 1 and No. 1A ESS in Chapter 7, except that its capacity is substantially less. Moreover, the ac input source is the usual 120-volt, 60-Hz, single-phase power. In addition, two regulated + 6.7-volt supplies are derived from the − 48 volt battery. The + 6.7-volt supplies are used for the logic units in the control complex of the No. 2 and No. 2B ESS.

Three power feeders, carrying − 48 volts, ground, and + 24 volts, run from the power plant to a power distributing frame. Battery feeders from this frame supply the individual equipment frames in the No. 2 ESS. Also, + 130 volts and − 130 volts are derived from the − 48-volt supply by dc-to-dc solid-state converters for coin control and other requirements. The layout of the power system generally follows that of the No. 1 and No. 1A ESS as depicted in Fig. 7-20.

The 20-Hz ringing voltage and the various tone frequencies are furnished by the ringing and tone frame. This frame provides duplicate 20-Hz ringing generators and the same frequencies as used in the No. 1 and No. 1A ESS for dial tone, audible ringing, busy-tone, and other tone frequencies, as shown in Table 7-1 for the No. 1 and No. 1A ESS. Both ac/dc and superimposed ringing are provided in the No. 2 and No. 2B ESS, for use on four- and eight-party lines.

No. 3 ESS for Rural Areas

Continued advancements in semiconductor technology coupled with reduced equipment costs made possible the development of the No. 3 ESS to serve rural areas by the Bell Telephone Laboratories. The No. 3 ESS has a capacity of approximately 4,500 lines, including two-, four-, and eight-party lines. It can handle 11,000 peak busy-hour calls. The No. 3 ESS requires only one-third of the floor space and uses considerably less power than step-by-step or other community dial (electromechanical) offices that serve rural localities. The extension of electronic switching service to small towns and rural communities provides improved and additional telephone services similar to that furnished in urban and suburban areas.

For instance, call forwarding, call waiting, abbreviated or speed calling, Touch-Tone dialing, and dial-tone first coin services become available. In addition, automatic number identification (ANI) also is furnished in the No. 3 ESS. Therefore, it is not necessary for operators to ask for customers' line numbers on long-distance calls.

No. 3 ESS Switching Network

Major elements of the No. 3 ESS are shown in Fig. 8-17. In rural localities, most calls are made between lines within the same central office. Thus, only a small number of interoffice trunks are needed. Therefore, only one type of a *switching network frame* is used in the No. 3 ESS for interconnecting customer lines, trunks, and service and junctor circuits. Furthermore, all interconnections are performed by three physical switching stages employing remreed crosspoint switches (see Fig. 7-6). Stages 1 and 2 are located on the switching network frame and the third stage is on the peripheral control frame, as illustrated in Fig. 8-17.

All calls (line-to-line, line-to-outgoing trunk, or incoming trunk-to-line) traverse five sets of remreed crosspoints. Calls pass through the first

Fig. 8-17. Major Elements of No. 3 ESS

and second stage switches in the switching network frame twice, and once through the remreed switch of the third stage in the peripheral control frame. Moreover, on every call, the connections between the second and third stages include a junctor circuit as one element, and a wire junctor, trunk, or service circuit as the other.

For example, referring to Fig. 8-17, an intraoffice call from customer (subscriber) A to B is first switched through stages 1 and 2 to the junctor circuit in the peripheral control frame. Next, it traverses the stage 3 remreed switch to the wire junctor (WJ), and then back through stages 2 and 1 to the line of customer B. Similarly, on an interoffice call, customer A is connected through stages 1 and 2 to the junctor circuit and stage 3. In this case, stage 3 switches the path from customer A to a trunk circuit that is routed back through stages 2 and 1 to the selected outgoing trunk C.

The trunk and junctor circuits provide supervision on the operations of their respective trunk and customer line circuits. They function in a similar fashion to those shown in Figs. 2-11 and 2-12 for the equivalent circuits in the No. 1 ESS. Service circuits include circuitry for receiving dialed digits and Touch-Tone signals, and supplying ringing current to called lines. Likewise, service circuits are used to provide access to tone and announcement circuits, thereby permitting such messages to be transmitted to customers with a minimum occupancy of the switching network.

In addition to switching stage 3, the peripheral control frame incorporates the junctor, trunk, and service circuits, their associated scanning circuits, and the peripheral pulse distributors with their related control circuits. The scanner circuits periodically check the state (on-hook or off-hook) of the above-mentioned circuits as directed by the on-line central control of the 3A processor. The peripheral pulse distributors transmit high-speed pulses, under the direction of central control, for the closure or release of switching network paths, and the control of junctor, trunk, and service circuits. These various operations are comparable to those previously described for equivalent hardware (equipment) units in the No. 2B ESS.

3A Processor Operations

The processor frame contains the two 3A processors (one on-line and the other on-standby). Each has its central control unit and main store memory as depicted in Fig. 8-17. Main store is a high-density memory in which are permanently stored call-processing and maintenance instructions, as well as all office records of customers, equipment and line data, and trunk connections. Status of calls in progress is temporarily stored in the call store memory section of main store.

Each 3A processor's central control unit, its main store, and one-half of the scanner control and peripheral distributor circuits in the peripheral control frame function as a *system control unit.* Consequently, if trouble should develop in any part of the on-line system control unit, it is immediately switched out of service. At the same time, the standby unit is placed on-line. this method saves a considerable amount of hardware and negates the need of fault-recovery software, thereby insuring greater service reliability.

Memory Functions in No. 3 ESS

Power interruptions or circuit faults can cause loss of the memory in main stores. Consequently, cartridge tape units are furnished on the maintenance frame to protect against this contingency. Furthermore, certain read-only permanent sets of instructions, called *microprograms,* are stored in the read-only memory (ROM) of central control. These instructions operate in tandem with the cartridge tape memory. Thus, in the event of a loss of memory in main stores, the microprogram will direct the cartridge tape unit to reload both main store memories. These operations correspond to those described for the 2B proccessor in the No. 2B ESS.

A prime advantage of the cartridge tape memory units is that it replaces the memory space in main stores for infrequently utilized low-priority programs. For example, recent programs cover changes in customer line numbers and their connections on the switching network, diagnostic procedures, and programs for the modifications of traffic and office data. These low-priority programs can be stored on cartridge tapes and, when required, immediately be supplied to main stores. Upon completion, the portion of the memory space that held the program in the cartridge tape is cleared for reuse.

Maintenance of No. 3 ESS

The maintenance frame, in addition to the aforementioned cartridge tape units, comprises the system status panel, the controllers for the cartridge tape units, and teletype terminal equipment for communicating with the No. 3 ESS as detailed in Fig. 8-17. Maintenance messages between the No. 3 ESS and its remote switching control center (SCC) normally are sent over dedicated data and teletype channels. In case of failure of the dedicated channel, maintenance channels are automatically routed over the DDD network to the distant SCC. Maintenance work also can be performed by means of the teletypewriter on the maintenance frame in the No. 3 ESS office.

The DDD network likewise is utilized for handling administration messages. These include traffic reports, line and trunk assignments to the switching network, and customer service data. Transmission of these mesages is controlled by the *autoconnect circuit* in the No. 3 ESS. This circuit "dials" the desired number, such as repair, traffic, or service order bureau (see Fig. 7-18). Upon receiving the call, a data set in the called bureau turns on its teleprinter (teletypewriter), which prints out the message. The connection is then released and the teleprinter and autoconnect circuit are available for other usage if no further messages are received after a fixed time interval.

Craftspersons in the above-mentioned bureaus desiring to communicate with the No. 3 ESS dial a specified number. The No. 3 ESS first verifies the request for the autoconnect circuit and calls back on another line to establish the connection. This procedure substantially reduces costs by employing the DDD network, and it also enables one teleprinter to serve a number of No. 3 ESS offices.

Error-analysis and fault-recovery programs in the No. 3 ESS automatically check for troubles in the switching network paths, trunks, and junctor and service circuits. Any defective equipment is switched out of service. If these programs are unable to restore system operations, the SCC is immediately notified via the aforementioned data and teletype channels. In this connection, SCC personnel can manually direct the No. 3 ESS to make necessary diagnostic tests, and to control operations of portions of the system as required to restore system operations. A duplicate of the system status panel (installed on the maintenance frame) is provided in the SCC for this purpose. Subsequently, craftspersons can be sent to the No. 3 ESS office to effect necessary repairs.

No. 3 ESS Test Frame and Power Plant

A *test frame,* usually installed adjacent to the maintenance frame, makes continuity and other operational tests on customer lines, trunks, and service circuits. It functions in a like manner as the trunk test frame used for the No. 2 and No. 2B ESS. Other hardware items, including Touch-Tone receivers, announcement machines, and special trunk and service circuits are installed on a miscellaneous frame. Equipment frames in the No. 3 ESS employ modular type, plug-in circuit packs, and connector-equipped cables. Hence, installation work is simple and the cost of future expansion and modifications is greatly lowered.

The power plant for the No. 3 ESS is analogous in many ways to the one in the No. 2B ESS. It furnishes standard 48-volt dc supply from the commercial 120-volt, 60-Hertz, single-phase ac mains. Dc-to-dc converters are used to derive the other dc voltage requirements. In the event of a commercial power failure, storage batteries provide the power for the No. 3 ESS, and the distant SCC is informed of the situation. Duplicate 20-Hz ringing generators, dial tone, audible ringing, busy-tone, and other needed tone frequencies are furnished similar to the No. 2B ESS, as listed in Table 7-1. Also, superimposed ringing is provided for use on four- and eight-party lines.

No. 1 EAX (Electronic Automatic Exchange)

The No. 1 Electronic Automatic Exchange (EAX) is a stored program and space division type of electronic switching system developed by GTE Automatic Electric Laboratories. It is designed to function as a Class 5 (or end central office), and as a combined Class 4 and 5 (toll and end central office). Generally, the No. 1 EAX would serve large urban localities. The capacity of the No. 1 EAX is 30,000 lines with medium traffic and 64,000 directory numbers; it can grow up to 44,000 lines. Its traffic handling capability is 79,000 busy-hour calls. The types of customer services and other features provided by the No. 1 EAX are essentially the same as those furnished by the No. 1 ESS and described in Chapter 2.

The No. 1 EAX system may be divided into three major groups: the switching network, the common control subsystems, and the maintenance control center. The switching network includes the line group, trunk register group, and the selector group frames. The common control subsystems are made up of the originating marker, terminating marker, register-senders, and the data processor unit. The maintenance control center (MCC) illustrated in Fig. 8-18 is the focal point for monitoring system operations. It serves as the interface between maintenance personnel and the subsystems of the No. 1 EAX. The maintenance control center also furnishes visual indications of traffic operations and the necessary controls for initiating test call routines and test programs. The teletypewriter provides printouts of maintenance information and related data.

The major groups and associated elements of the No. 1 EAX and their correlations are illustrated in the block diagram of Fig. 8-19. Each line group frame provides for 1,000 customer lines through four stages of switching (used for different purposes), designated A, B, C, and R. Switching stages A and B are used for originating calls, such as from customer line (1). These switches select a path from the originating line to a spare originating junctor. Switching stage R connects the selected originating junctor to one of twenty originating register junctors in order to reach a register-sender for subsequent receiving and sending of dialed information. An incoming call would be routed through a terminating junctor to the A, B, and C switching stages in the line group frame in order to be connected to the called party, such as to customer line (2). The terminating junctors provide ringing control, talking battery for both parties, and call supervision. Both the originating and terminating junctors remain in the transmission path for the duration of a call.

Fig. 8-18. No. 1 EAX Maintenance Control Center (Courtesy of GTE Automatic Electric Co.)

Three switching stages are provided in the selector group network as shown in Fig. 8-19. They provide interconnections between originating junctors and outgoing trunks and between incoming trunks and terminating junctors or outgoing trunks. The trunk register group is equipped with two switching stages. It furnishes a temporary signaling path between 200 incoming trunks and forty incoming register junctors per group. The trunk register group and its associated register junctor provide access to an idle register-sender. These equipment units are utilized only during the establishment of an incoming call in order to record the called line number in the register-sender for subsequent processing.

The actual operation of the crosspoint connections in the switching network are performed by *correeds*. The correed consists of two reedcapsules surrounded by a coil winding. Each reedcapsule is composed of two magnetic reeds equipped with gold-plated contacts that are hermetically sealed in a glass enclosure or capsule. The reed contacts in one reedcapsule are used to switch the tip conductors, and the reed contacts in the other reedcapsule switch the ring conductors of the particular talking path involved.

The correeds used in No. 1 EAX installations are comparable in construction and operation to the ferreed crosspoint switches initially employed in the Bell System's No. 1 ESS. Unlike the ferreed, however, it is necessary that a small direct current continuously flow through the windings of the correed in order that its reed contacts can operate and remain closed. The

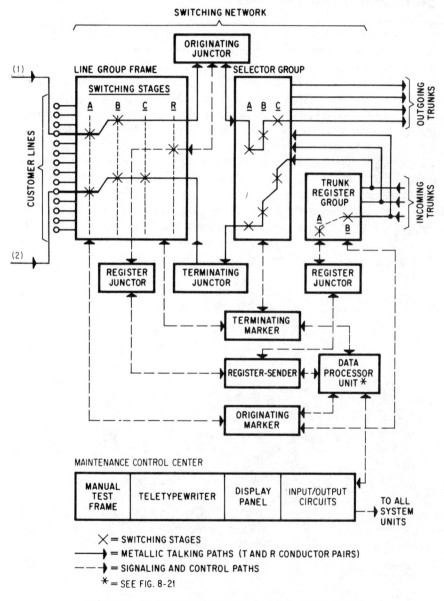

Fig. 8-19. Correlations of Major Elements of No. 1 EAX

reed contacts open when the current flow ceases. A number of correeds are mounted on a printed wiring card to form a matrix for use in switching stages of the No. 1 EAX. Figure 8-20 shows a typical correed matrix.

Call Control Operations in No. 1 EAX

The register-sender is a time shared multiplex and common control unit. It has wired logic and memory facilities for receiving and storing dialed digits from originating customer lines, as well as the directory number of the called line on incoming calls. Approximately 390 simultaneous originating and terminating calls can be handled. In addition, the register-sender generates pulses for transmitting outgoing call data via outgoing trunks. A temporary ferrite-core memory is utilized by the register-sender for this purpose in much the same manner as call store in the No. 2 ESS. Similarly, the common logic and memory units in the register-sender are in duplicate and are operated in synchronism to provide instantaneous error and fault detection facilities.

The originating marker detects originating calls from customer lines in the line group frames, and terminating calls from incoming trunks in the

Fig. 8-20. Correed Matrix Used in No. 1 EAX (Courtesy of GTE Automatic Electric Co.)

trunk register group. It controls path selection between the line group and the originating junctor, and between the originating junctor and the register junctor as indicated in Fig. 8-19. The originating marker also controls the selection of paths between incoming trunks connected to the trunk register group and register junctors. These functions may be likened to those of the originating marker in the No. 1 Crossbar office as explained in Chapter 1 and Fig. 1-10. The originating marker also provides the data processor unit with the identities of the calling line and the originating and register junctors involved, and executes the orders received from the data processor unit for completing the originating call process.

The terminating marker directs the establishment of paths for all calls through the selector group. It also sets up paths through the line group to the called line as represented in Fig. 8-19 by the paths to called line (2). The terminating marker functions under the control of the data processor unit. For instance, the terminating marker selects idle paths, either randomly or sequentially, from the specified group, as directed by the stored program in the data processor unit.

Both the originating and terminating markers are installed in duplicate and both operate in pairs. Each can process separate calls simultaneously unless a fault should occur. In this event, the malfunctioning unit would be automatically taken out-of-service.

Organization of No. 1 EAX Data Processor Unit

The data processor unit, illustrated in Fig. 8-19, directs call processing, maintenance, and related operations in the No. 1 EAX. Its organization and principal components are represented in block diagrams in Fig. 8-21. The central processor unit (CPU) with its main memory and magnetic drum memory systems, and the respective memory control circuits, make up the computer complex of the No. 1 EAX.

The CPU is essentially a general-purpose computer employing large-scale integrated (LSI) circuits. Under the control of stored programs in main and magnetic drum memory systems, it governs various call-processing, system diagnostics, maintenance, and other pertinent functions. Effective commands for required system operations are sent to peripheral and system units by means of input/output devices in the data processor unit shown in Fig. 8-21. These input/output devices include: communication registers used with the originating and terminating markers, and the input/output service buffers associated with data sets and teletypewriters. Two other input/output equipments, the ticket scanner and magnetic tape controllers, work in conjunction with the stored program in the main and drum memories to provide automatic message accounting (AMA) service.

Fig. 8-21. No. 1 EAX Data Processor Unit Organization

The register-senders, originating and terminating markers, and the maintenance control center are directly associated with the data processor unit (see Fig. 8-19) for the transmission of information and the reception of operational instructions. Communication links are furnished by peripheral buses to connect with the computer channel multiplexer. In this way, central processor (CPU) is provided with information regarding the processing of calls by subsystems, status of subsystems, and data concerning the input/output devices.

The main memory is of the random-access type that stores instructions and data that are frequently utilized by the CPU. In the initial No. 1 EAX installations, the main memory employed ferrite-core devices to store programs for its ROM and read/write modes of operation. The capacity of the ferrite-core memory is 32K (32,768) binary words, each of 26 bits (including 1 parity and 1 protect bit). Subsequent technological improvements led to the development of metal-oxide semiconductors (MOS) in place of ferrite cores for the main memory. The MOS main memory has a maximum capacity of 128K (131,072) binary words. (Note that K = 1,024 binary digits or bits.)

The *drum memory* system serves as a storage area for programs and data that are not needed for immediate access by central processor (CPU). Information is read out of the drum memory and written into main memory, via their respective control units (see Fig. 8-21), as required by central processor. Each drum memory unit consists of a metallic cylinder, approximately 10 inches (25.4 cm) in diameter and 28 inches (71 cm) long, which is motor-driven at about 3,600 rpm. The cylinder's surface is coated with nickel-cobalt magnetic material for recording binary data.

The magnetic drum is divided into 18 segments for storing binary data, and 11,000 locations are provided around its periphery for this purpose. Thus, each segment can record and store 11,000 binary words, each of 27 bits. In addition, every segment is accessed by 28 recording heads (including one spare). The storage capacity of a magnetic drum, therefore, is 198,000 binary words. And there are a total of 510 recording heads (18 segments × 28 = 504 plus 6 clock-track recording heads). Depending on the size of the No. 1 EAX installation, from four to a maximum of six magnetic drum memories are provided.

The generic program normally used in the No. 1 EAX consists of approximately 197,000 binary words, each of 27 bits. The magnetic drum memories usually contain 160,000 binary words and the remainder (37,000 words) are stored in main memory. More than half of the stored program instructions are concerned with routine system maintenance and diagnosing faults. Data stored in the main and drum memories depend on the number of lines and trunks equipped in the particular No. 1 EAX office.

Four major programs, each of which is capable of performing one or more system functions, make up the software in the No. 1 EAX. These programs are designated: Operating System, Telephone Applications, Diagnostics, and Office Administration.

The *operating system programs* direct the detailed interface operations required by the other programs for control of the various input-output devices and subsystems. They may be compared to the executive program in the No. 1 and No. 1A ESS because it also schedules the tasks to be performed by the data processor unit. In addition, operating system programs provide many common functions, such as the allocation of buffers and the timing of operations.

Telephone application programs contain the control logic and data processing needed to furnish the basic services and features of an electronic switching central office. These programs cover call-processing, ticketing, and metering groups. The call-processing program directs the various functions for processing a call within the No. 1 EAX. For example, it determines the class of service, analyzes the dialed digits, initiates path and route selections, and decides call terminations. The main memory is used for translations and other data retrievals related to the particular call in progress. The metering program records various system conditions detected by the call-processing programs. Automatic message accounting (AMA) operations are directed by the ticketing programs, which also work in conjunction with the call-processing programs. Ticketing programs determine the answer and disconnect time of calls that are to be ticketed for charge purposes, so that this information can be recorded by the magnetic tape unit.

Diagnostic programs, as their name implies, are concerned with system diagnostics and maintenance operations. For instance, errors and faults detected by the various subsystem circuits are conveyed to the diagnostic programs by fault interrupts. These programs endeavor to locate and isolate the faulty unit, and to identify and localize the fault condition. The type of fault and its location are then communicated to the maintenance personnel by means of a data set and teletypewriter in the maintenance control center. Diagnostic programs also provide for routine checks of line, trunk, and junctor equipments.

Office administration programs make major and minor changes to the stored program. They also serve to add, delete, or modify data used in the processing of a call.

No. 2 EAX (Electronic Automatic Exchange)

The No. 2 EAX, also developed by GTE Automatic Electric Laboratories, is a stored program control and space division analog electronic switching system. It is designed for use as a Class 5 or end central office but can provide local tandem switching as part of its end-office functions. Although intended for initial installations of about 2,000 lines in suburban areas, it can be expanded to 40,000 lines with 4,000 inlets for trunks. Its traffic-handling capacity is 60,000 completed busy-hour calls. Customer services furnished by the No. 2 EAX are equivalent to those of the No. 1 EAX. In addition, Centrex service can be provided for business customers.

The equipment (hardware) of the No. 2 EAX may be divided into two major functional categories: the *peripheral and switching network* and the *2A processor complex,* as illustrated by block diagrams in Fig. 8-22. The switching network is a two-wire folded six-stage type. It has a maximum capacity of 16 switching network units. Each network unit has a capability of 4,096 inlet and 1,024 outlet terminations. The inlet terminals connect to customer line circuits, service and trunk circuits, and the maintenance (Mtce.) and test circuit. The outlet side of each switching network unit con-

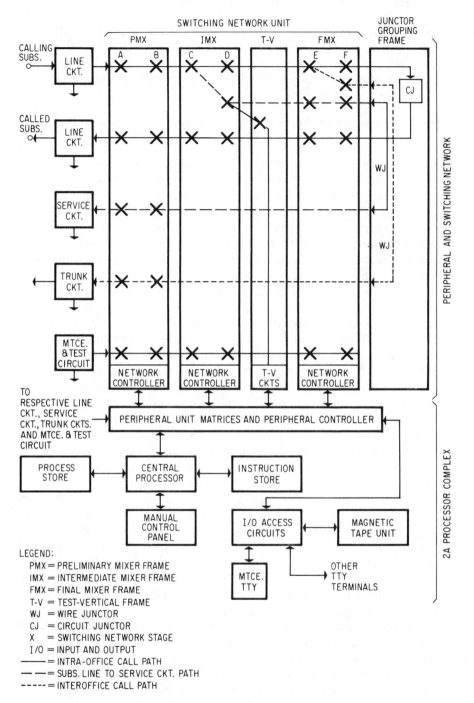

LEGEND:
PMX = PRELIMINARY MIXER FRAME
IMX = INTERMEDIATE MIXER FRAME
FMX = FINAL MIXER FRAME
T-V = TEST-VERTICAL FRAME
WJ = WIRE JUNCTOR
CJ = CIRCUIT JUNCTOR
X = SWITCHING NETWORK STAGE
I/O = INPUT AND OUTPUT
———— = INTRA-OFFICE CALL PATH
— — = SUBS. LINE TO SERVICE CKT. PATH
----- = INTEROFFICE CALL PATH

Fig. 8-22. Major Equipment of No. 2 EAX

261

nects to either *wire junctors* or *circuit junctors.* Wire junctors provide direct metallic connections for the paths in the switching network between customer line circuits and service circuits, such as, for receiving dial pulses from a calling customer. Circuit junctors contain supervisory and transmission devices for use on intraoffice calls.

2A Processor Complex of No. 2 EAX

The major elements of the 2A processor complex are: central processor, instruction store and process store memory units, the peripheral controller with its peripheral unit matrices, input/output access circuits, and the manual control console. These components likewise are shown in the block diagrams of Fig. 8-22. The peripheral unit matrices of the peripheral controller serve as the communication interface between the very high-speed central processor and the slower-speed equipment in the peripheral and switching network. Moreover, with the exception of the input/output access equipment and the manual control console, the other main elements of the 2A processor complex are duplicated and also are interconnected by a bilateral bus system for maximum reliability.

Call-processing and maintenance operations in the No. 2 EAX are directed by the duplicated central processor in the 2A processor complex. Both central processors normally operate in synchronism with each other; one serving as the active unit and the other as standby equipment. Each central processor performs the call-processing tasks in step with the other so that, in the event of a failure, the standby unit can immediately assume control. The central processor essentially is a computer. Under the direction of programs in instruction store, it acts on data presented by process store and peripheral controllers to handle various electronic processing and switching operations. Therefore, the central processor functions in a similar manner as is described in Chapter 4 for central control in the No. 1 ESS.

The memory system in the No. 2 EAX consists of the instruction store and process store units also indicated in Fig. 8-22. Both memory units utilize identical types of semiconductor integrated circuits (IC) with a capacity of 512,000 binary words. Each binary word comprises 32 bits plus 1 parity bit for error-checking purposes. Instruction store has a protected memory containing the program instructions for use by the central processor in handling calls. The memory in instruction store cannot be modified in normal call-processing operations.

Process store includes both protected and unprotected (erasable) memories for use by the central processor. Thus, the central processor may write binary data into process store for temporary storage until it is read out as needed during the progress of a call. The functions of instruction store and process store are comparable to those of program store and call store, respectively, in the No. 1 ESS, as explained in Chapter 5.

The input/output access frame enables central processor to access and control the magnetic tape unit, teletype (TTY), and data set terminal facilities, which are illustrated in Fig. 8-22. The magnetic tape unit stores data recorded on calls handled for subsequent use in billing customers, and for maintenance and traffic study requirements. Likewise, the magnetic tape unit stores memory changes for instruction store and process store. The data set and TTY facilities provide communication channels between the No. 2 EAX (via the maintenance control console) and maintenance personnel at local and remote locations. These communication channels can be considered comparable to those furnished in the No. 2 and No. 3 ESS, previously described.

The control and communication interface between the No. 2 EAX and maintenance personnel is the maintenance control console depicted in the block diagrams of Fig. 8-22. Its control panel furnishes craftspersons with continuous visual indications of the status of the 2A processor and provides the means to manually select and hold desired configurations of subsystems in the 2A processor. In addition, the maintenance control console activates visual and audible alarms in case of equipment or software malfunctions and, when necessary, initiates the loading of programs from the magnetic tape unit into the memory circuits of instruction store and process store. Furthermore, it provides means for the manual control of system operations, or to inhibit certain system maintenance actions in emergencies.

2B Processor for No. 2 EAX

An improved stored-program-controlled processing system, designated the *2B processor,* has been developed by GTE Automatic Electric Laboratories. It is a direct replacement for the 2A processor and also will be used in the No. 3 EAX in connection with digital switching operations. The 2B processor can complete call-processing and maintenance software functions at much greater speeds than the 2A processor. This higher execution speed permits the future additions of new service features and results in a greater call-handling capacity of the switching network of the No. 2 EAX.

Improved semiconductor devices, particularly large-scale integrated (LSI) circuits, are utilized to attain the higher processing speeds in the central processor of the 2B processor. Moreover, microprograms (also termed firmware) are stored in the read-only memory (ROM) system of the 2B processor. These microprograms enable the 2B processor to work with the program instructions of the 2A processor. Additional instructions may be added to the 2B processor by only changing information contained in the read-only memories (ROMs). The ROMs of the central processor unit include microprograms for the maintenance functions of the 2B processor as well as microprograms to reconfigure the 2B processor after a malfunction has been detected by the equipment.

Switching Network of No. 2 EAX

The switching network of the No. 2 EAX contains six switching stages, designated AB, CD, and EF, and a test-vertical (T-V) frame as per the block diagrams in Fig. 8-22. The basic switching element in each stage is the grid card. It consists of a printed-wire card containing eight 4 × 4 matrix arrays of a new correed-type magnetic latching reed (MLR) crosspoint developed by GTE Automatic Electric Laboratories. Figure 8-23 is a schematic diagram of this grid card. This drawing shows some of the connections provided by the 128 MLR crosspoints between the 16 inlets and the 16 outlets. Note that each inlet and outlet constitutes a tip and ring pair.

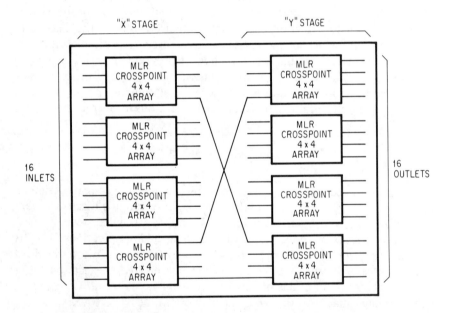

Fig. 8-23. Grid Card with 4 x 4 Arrays of MLR Crosspoints Used in No. 2 EAX (Courtesy of GTE Automatic Electric Labs.)

The magnetic latching reed-capsule (MLR) is an improved correed type that is similar in appearance but much smaller in size than the correed used in the No. 1 EAX. It employs a Remendur alloy material for the reed, which has rhodium-plated contacts. However, a continuous flow of direct current is not required for operation as for the correeds in the No. 1 EAX. The MLR closes its contacts upon the application of a current pulse (approximately 5.5 amperes) for a duration of 1.5 milliseconds through both of its horizontal and vertical windings. A current pulse through either winding will reset the MLR and open its contacts. Thus, the MLR may be compared to the remreed crosspoint (see Fig. 7-6) used in the No. 1A ESS.

The test-vertical circuits on the test-vertical frame act as links between switching stages D and E (see Fig. 8-22). These circuits, as directed by the central processor, are used to check for foreign potential, crosses, or false grounds on selected transmission paths in the switching network. These functions are similar to these performed by the test verticals and their FCG and no-test circuits that are utilized to check switching network paths in the No. 2B ESS, as illustrated in Fig. 8-15.

Call Operations in No. 2 EAX

Switching network paths used for intraoffice and interoffice calls are illustrated in Fig. 8-22 by their respective (————) and (—·—·—) lines. The dash lines (————) indicate network paths from a calling customer's (subs.) line circuit to a service circuit, which is set up when a call is originated. The following is a synopsis of the various connections and processing steps usually involved for an intraoffice call within the No. 2 EAX.

The off-hook state of the calling line is first detected during periodic scannings of its line circuit. This information is sent back to central processor, through the particular peripheral matrix associated with the line circuit and the peripheral controller, as may be perceived in Fig. 8-22. Central processor causes the calling line's equipment number (not the directory number) to be stored in a call register of the unprotected (temporary) memory section of process store. Concurrently, an idle dial pulse receiver (DPR) service circuit is selected and its associated circuit register, in the unprotected memory part of process store, is made busy. The memory location or address data of this DPR service circuit is also recorded in the aforementioned call register (which had previously stored the equipment number of the calling line. Next, central processor directs a search of the software containing the switching network map (refer to Fig. 6-7 and its related text) in process store for an idle wire junctor and network paths to it. One path is from the calling line circuit and the other is to the previously selected DPR service circuit. The data recorded previously in the above-mentioned call register and circuit register determine the paths selected. These paths are represented by the dash lines (————) in Fig. 8-22 drawn from switching stage C to the *DPR service circuit*. Central processor now directs the closing of these paths in the switching network so that dial tone can be sent to the calling customer from the DPR service circuit.

The called line number, dialed by the calling customer, is recorded by the DPR service circuit for subsequent processing. Central processor, under control of programs in instruction store, initiates the processing actions. The called line number is first translated into its line equipment number. This equipment number is then checked for idle or busy status in the temporary memory of process store. If found idle, the called line circuit is made busy. The network map is next examined for an idle circuit junctor and two paths to it; one from each of the respective calling and called line circuits. However, only the path to the calling line circuit is closed at this time. The other path to the called line circuit is reserved but not closed through.

Simultaneously, an idle ringing current service circuit and another wire junctor are selected together with two network paths to this wire junctor. One path from the ringing current service circuit and the other from the called line circuit are needed. Ringing current is now applied to the called line via the aforementioned paths and wire junctor. When the called customer answers, the off-hook condition is detected by the above-mentioned service circuit, and central processor is informed. Thus, central processor directs the release of the ringing current service circuit, its network paths, and the associated wire junctor. And the reserved path connection from the circuit junctor to the called line circuit is now closed through. Transmission battery and supervision over the calling and called lines are provided by the circuit junctor until disconnection of the call.

The reader is referred to Fig. 6-7 and its related sections, "Network Map" and "Functions of Path Memory on Intraoffice Calls," which describe call operations in the No. 1 ESS. These explanations are applicable in many respects to the foregoing operations in the No. 2 EAX. In this connection, note that call store in the No. 1 ESS may be considered comparable to process store in the No. 2 EAX.

Interoffice calls in the No. 2 EAX are handled in a similar fashion as described for intraoffice calls, except that a circuit junctor is not required. Instead, a wire junctor is selected to link the calling line circuit to an idle trunk circuit in order to connect to the terminating central office. The paths for an interoffice call in the switching network are illustrated by the (—·—·—) lines, from switching Stage E to the wire junctor and then to the trunk circuit, portrayed in Fig. 8-22. The trunk circuit furnishes transmission battery to the calling customer and transmission and signaling means to the distant central office equipment.

Questions

1. Why is it not feasible to utilize the No. 1 ESS and No. 1A ESS to serve communities in suburban and rural areas? What types of Bell System electronic switching equipment have been developed for these localities?

2. Name two or more major electronic telephone switching systems, in addition to the Bell System types, and indicate their manufacturers.
3. What comprises the control complex of the No. 2 ESS? What common unit is associated with it? What equipment unit may be considered equivalent to central control in the No. 1 ESS?
4. Name the major equipment frames in the No. 2 and No. 2B ESS.
5. Name the major elements of the 2B processor. How does its size compare to the processor in the No. 2 ESS?
6. Describe the principal functions of the maintenance frame in the No. 2B ESS.
7. How many major frames and switching stages are provided in the No. 3 ESS? What are microprograms?
8. What three main equipment groups make up the No. 1 EAX? Briefly describe the two memory systems provided in the No. 1 EAX.
9. Name the two principal equipment categories of the No. 2 EAX. What type of switching network is utilized? Describe the basic switching element used in each switching stage. How does it differ from that used in the No. 1 EAX?
10. What are the main parts of the 2A processor complex in the No. 2 EAX? How does the 2B processor attain its higher processing speeds?

9
Digital Telephony

Analog and Digital Signals' Fundamentals

Speech and data signals originating at telephone stations are in *analog* form. In other words, these signals vary continuously in voltage or current magnitude, corresponding to the amplitude variations of the signal's waveform with time. The electromechanical and electronic switching systems previously described in this book are designed to switch analog signals. These analog switching systems commonly employ space-division switching technology. This necessitates that a two-wire metallic path be established for every connection through the switching network.

Analog voice signals may be changed to digital form by means of the sampling technique. The resultant digital signals represent a series of sampling pulses, each proportional to the amplitude of the analog signal's waveform (usually its voltage wave) and corresponding in time to it. Figure 9-1 illustrates this time-division relationship which is appropriately termed pulse amplitude modulation (PAM).

To faithfully transmit and subsequently reproduce the original speech signal, it is essential that it be sampled at a rate at least twice its highest significant frequency. Thus, the 8,000-Hz sampling rate has been standardized as the initial step in converting the nominal 4-kHz bandwidth analog channels (200–3400-Hz voice-frequency signals) to a digital format. Each PAM sample (see Fig. 9-1) is next quantized and encoded into an eight-binary digit (bit) word of a pulse-code modulation (PCM) signal. One bit is used as the polarity indicator and the other seven represent the sample's amplitudes. These seven bits are quantized or compared electronically to a binary-code scale of 255 values. Then, they are encoded with the polarity-indicator bit to form PCM digital signals as illustrated in Fig. 9-2.

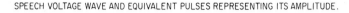
SPEECH VOLTAGE WAVE AND EQUIVALENT PULSES REPRESENTING ITS AMPLITUDE.

SPEECH VOLTAGE   PULSE AMPLITUDES

Fig. 9-1. Analog Vs. Digital Signals

PCM Digital Signals

Digital signals require four-wire transmission paths (two cable pairs or two separate paths in a switching network), one for each direction of transmission. Digital signals, however, have a number of advantages over analog signals. They will not be degraded by noise, crosstalk, or distortion during transmissions as in the case of analog signals. Moreover, digital signals, unlike analog, can be regenerated in amplifiers or by repeaters installed at regular intervals along cable routes.

While it may not seem efficient for PCM carrier systems to send a number of pulses, instead of one pulse, for each sample of a voice signal,

NOTES:

1. SIGNALING STATE PULSE ONLY SENT IN THE 8th TIME-SLOT OF FRAME NOS. 6, 12, 18 AND 24.

2. ONE BIT FOR SYNCHRONIZING AND FRAMING IS ADDED AT THE END OF CHANNEL NO. 24.

Fig. 9-2. Typical PCM Signals of T1 Carrier System

this technique is an advantage, because it is only necessary, at any given time, to detect whether or not a pulse is sent. This requirement is less difficult to accomplish than to determine the amplitude of a pulse. Consequently, various types of information (voice, data, facsimile, and video) can be multiplexed for PCM transmission and also for subsequent digital switching operations.

Digital telephone carrier systems employing PCM techniques were first developed by the Bell Telephone Laboratories. The Bell System's T1 carrier systems are used extensively to provide high-quality interoffice, tandem, and toll-connecting trunks at low cost. The T1 PCM carrier multiplexes 24 *analog* voice channels into *digital* signals utilizing a sampling frequency of 8,000 Hz. These signals are then quantized and encoded into an eight-bit nonlinear PCM signal of 193 bits (24 × 8 plus one bit added for synchronization). The resultant PCM 24-channel bank is designated a *digroup.* The original digroup, type D1, has been replaced by type D3, which provides improved transmission quality required for the Bell System's DDD or intertoll network. The sampling, quantizing, and encoding processes for all 24 channels are completed within 125 microseconds (1/8,000 Hz). This results in a transmission rate of 1.544 Megabits per second (Mb/s) (193 bits/125 microseconds), which has been standardized for T1 PCM carriers.

(The foregoing descriptions of PCM carrier systems are extracted from the author's *Basic Carrier Telephony,* published by Hayden Book Company, Inc., in 1977, to which the reader is referred for further explanations of PCM system operations.)

Digital Switching Concepts

Digital voice signals may be switched by both space- and time-division processes. For example, crossbar and remreed switches can provide metallic space-division switching, and time-division switching can be performed by time-division multiplex (TDM) electronic means. In both methods, operations of digital switching systems are governed by the same stored program control (SPC) techniques that are used in analog electronic switching offices (such as No. 1 and No. 1A ESS) described in previous chapters.

An early type TDM digital switching system is briefly described in the first section of Chapter 2 and illustrated in Fig. 2-1. It was designed primarily for PBX installations and small central offices because of its inability to efficiently handle the high calling rate and increased traffic volume needs of large electronic switching offices.

Subsequent advancements in semiconductor technology produced low-cost and multifunction large-scale integrated (LSI) circuits or chips. These chips were instrumental in the subsequent development of economical analog to digital (A/D) and digital to analog (D/A) converters, as well as very high-speed logic gates. Thus, incoming analog signals could be readily changed to the PCM format for digital switching operations. The PCM signals are then converted back to analog for transmission over analog carrier systems or to analog central offices. Digital signals from or to PCM carrier systems do not need this conversion treatment, thereby further reducing equipment costs.

The rapid growth in the use of PCM T1 carrier systems for interoffice and toll trunks was an important factor in Bell Telephone Laboratories developing in 1976 the first toll digital switching system, termed No. 4 ESS. Digital switching for central offices serving customer lines evolved later. Furthermore, four-wire transmission also was commonly employed by analog toll trunks, including those derived from frequency-division multiplex (FDM) cable carrier and microwave systems. Thus, A/D and D/A conversions could be satisfactorily accomplished without providing hybrid transformers and their associated balancing networks needed for two-wire to four-wire terminations. Figure 9-3 is a schematic diagram of a typical hybrid circuit for this purpose.

The hybrid transformer has three pairs of balanced coils, designated primary (p), secondary (s), and tertiary (t) as depicted in Fig. 9-3. These

LEGEND:

LPF = LOW-PASS FILTER
A/D = ANALOG TO DIGITAL CONVERTER AND ENCODER
D/A = DIGITAL TO ANALOG CONVERTER AND DECODER
(p) = PRIMARY COIL OF HYBRID TRANSFORMER
(s) = SECONDARY COIL OF HYBRID TRANSFORMER
(t) = TERTIARY COIL OF HYBRID TRANSFORMER

Fig. 9-3. Hybrid Circuit for Two-Wire/Four-Wire Analog Trunk

coils are mounted on a common permalloy-type core. Incoming voice signals from the two-wire analog trunk (leads T&R) are transformer-coupled
(coils p to s) to path T_2,R_2 for A/D conversion and digital switching operations. Because of the balanced relationship of these three coils, only a very
small amount of the incoming voice signals are induced into the opposite
transmission path T_1,R_1. Similarly, the analog signals arriving on the T_1,R_1
path from the called party are sent back only over the T&R leads of the two-
wire toll trunk to the calling party. The *balance network* affords a proper
impedance match for this balancing of the various induced voltages.

The balance network of the hybrid arrangement, shown in Fig. 9-3,
simulates the impedance or electrical characteristics of the two-wire analog
toll trunk. The impedance of different trunks (nominally 600 ohms) is fairly
constant. Therefore, the impedance of the balance network does not have to
be changed when different trunks are connected. The impedance of customer lines, however, depends on the wire gauge, cable characteristics, and
the distance to the central office. Each line, therefore, could have a different impedance than the nominal design value of 900 ohms. Consequently,
an adjustable or compromise network would be required for the proper
digital switching of subscriber lines, to preclude feedback or crosstalk effects due to circuit imbalance. This situation was another reason for initially
providing digital switching to toll centers rather than to central offices.

For efficient digital switching, numerous time-division multiplex
(TDM) buses or lines are needed to handle the coded digital speech samples
(PCM signals). Each TDM line comprises a series of time-slots or positions
on the bus, one for each of the samples of a particular PCM voice channel
of a telephone call. To interconnect calls (from incoming to outgoing
trunks, or from calling to called telephone customers) necessitates that a
particular time-slot on one TDM line (bus) be immediately connected to the
designated time-slot in the same or another TDM bus. This process, appropriately called time-slot interchange (TSI), is utilized in the No. 4 ESS and
other digital switching systems. It also affords time buffering or delay to incoming digital pulses from toll trunks. This time-delay action is essential for
synchronizing these pulses with the timing of the switching network.

Both space- and time-division methods are employed for digital
switching of PCM channels connected to the respective TDM buses or lines.
In order to give the reader a better understanding of time-division switching, the following simplified analogy is presented: Let us first portray the
TDM buses as a series of vertical lines, and that every line connects to a 24-
channel PCM (T1 carrier) system. Next, assume that each line is divided
into 24 equal segments or time-slots, one for each of the 24-digital voice
channels of the PCM carrier. These representations are illustrated in Fig.
9-4.

Now, suppose that incoming toll trunk A (from a calling party) is to
be connected to outgoing toll trunk B for subsequent connection to a called

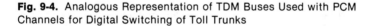

LEGEND:

− − − − − SPACE-DIVISION SWITCHING

• • • • • • • • TIME-DIVISION SWITCHING

Fig. 9-4. Analogous Representation of TDM Buses Used with PCM Channels for Digital Switching of Toll Trunks

party. Referring to Fig. 9-3, it is observed that four-wire switching is required for the two directions of transmission, that is, path T_2R_2 for the digitized speech signals from the calling to called party, and path T_1R_1 for the reverse called to calling party direction. It is further assumed that trunk A connects to channel 4 on bus 1 and that trunk B connects to channel 9 on bus 3. The following is a brief summary of the essential steps involved as depicted in Fig. 9-4.

1. Channel 4 on bus 1 (trunk A) is connected by space-division switching to the corresponding channel 4's time-slot position on bus 3.
2. The time-slot position of channel 4 on bus 3 is delayed (by means of time-division switching) for five channel time-slot periods to the time-slot period of channel 9 on bus 3. Therefore, trunk B (connected to channel 9 on bus 3) will be effectively connected to trunk A (channel 4 on bus 1) for the called to calling party transmission direction.

3. Simultaneously, channel 9 on bus 3 (trunk B) is likewise connected by space-division switching to the corresponding channel 9's time-slot position on bus 1.

4. The time-slot position of channel 9 on bus 1 is now delayed (by time-division switching) for a total of 19 channel time-slot periods to the time-slot position of channel 4 on bus 1 (24 PCM channels minus 5 channel time-slot periods = 19). This delaying action permits trunk A (channel 4 on bus 1) to be connected to trunk B (channel 9 on bus 3) for the calling to called party transmission direction.

The foregoing digital switching operations are performed by high-speed crosspoints and logic gates in semiconductor chips under the direction of a central processor and stored-program instructions, such as, central control and program and call stores in the 1A processor of the No. 4 ESS. Thus, the trunk A to trunk B connections are repeatedly established for the repetitious transfer of PCM signals in each direction of transmission. This repetitive action is governed autonomously by the time-slot interchange (TSI) and time-multiplex switch (TMS) units. The calling and called parties are only connected to each other for approximately 1 microsecond every 125 microseconds of the PCM-processing period. The speech transmission quality, however, is comparable to that obtained with a continuous connection.

No. 4 ESS Toll and Tandem Digital Switching System

The placing into service by the Bell System of the first No. 4 ESS in 1976 accelerated the trend to digital switching. It started with the digital switching of toll and tandem trunks in toll centers (Class 4 offices) and subsequently led to digital switching of customer lines in end (Class 5) central offices (see Fig. 1-2). The No. 4 ESS, developed by the Bell Telephone Laboratories, is a highly sophisticated time-division digital switching system. It has a maximum capacity of 107,520 trunks (four-wire circuits) and can switch 550,000 calls in the peak busy hour. The No. 4 ESS can process four times as many calls as the analog No. 4A crossbar system, which it replaced. Moreover, the No. 4 ESS requires only one-quarter of the floor space and has five times the terminal capacity of the analog No. 4A electro-mechanical system.

Basic Organization of No. 4 ESS

The principal elements and main peripheral units of the latest type No. 4 ESS are shown by simplified block diagrams in Fig. 9-5. The main features and functions of these components are outlined or further explained below.

Fig. 9-5. Principal Elements of No. 4 ESS (Courtesy of Bell Telephone Laboratories)

LEGEND:

CCIS = COMMON CONTROL INTEROFFICE SIGNALING
MAC = MACHINE ADMIN. CENTER
MCC = MASTER CONTROL CENTER
MOC = MACHINE OPERATIONS CENTER
TEC = TERMINAL EQPT. CENTER
TMS = TIME-MULTIPLEXED SWITCH
TSI = TIME-SLOT INTERCHANGE
TTY = DATSPEED #40 TERMINAL OR EQUIVALENT
⊖ = COAXIAL CABLES

Incoming and outgoing T1 digital trunks (each comprising 24-PCM voice channels also termed DS-1 signals) terminate directly on the digital interface frame (DIF). The DIF serves as the transmission interface with the time-division digital switching network. The interconnections of incoming toll trunks to outgoing trunks by the digital switching network are controlled by the 1A processor as subsequently explained.

Analog *frequency-division multiplex (FDM)* carrier systems (usually transmitted over coaxial cables) connect to the LT-1 connector frame. The LT-1 functions as a *transmultiplexer* becauses it changes incoming analog FDM carrier signals into the digital format for termination in the DIF. Similarly, it converts digital signals from the DIF back to FDM form for transmission by outgoing analog FDM carrier systems.

Analog metallic trunks (two-wire and four-wire) connect to the D4 channel bank equipment. This equipment converts the analog signals to the PCM digital form for multiplex operations in the DIF. The D4 channel banks, likewise, change the digital signals back to analog for transmission over individual outgoing metallic trunks and service circuits.

Digital Echo Suppressor Terminal For No. 4 ESS

Analog trunks are susceptible to talker *echo effects.* The echos are usually initiated by the reflection back to the respective talker of some of the calling or called party's voice signals. The reflections result from varying impedance characteristics that are encountered at interconnection terminals, for instance, at the hybrid interface of two-wire/four-wire terminations as depicted in Fig. 9-3. Because toll cables are many miles in length, the propagation speed of the returned echo signal is often delayed many milliseconds. This delay causes an annoying echo effect to the talker.

To suppress these undesirable echo effects, the respective return transmission path is effectively short-circuited only during the time that the calling or called party is talking. In this way, the particular talker will not hear his or her echo. Considerable electromechanical equipment was required for echo suppression of individual analog toll trunks and the cost was substantial.

Recent advances in semiconductor technology made feasible the development of a digital *echo supressor terminal (EST)* by the Bell Telephone Laboratories for use in the No. 4 ESS. It comprises digital common control echo suppressor units which function with the PCM signals transmitted between the digital interface frame (DIF) and the digital switching network. Echo suppressor operations in the EST are controlled by the 1A processor over the peripheral unit bus (PUB). Central Control in the 1A processor causes the EST to connect and disconnect echo suppressors, as required, on the individual toll trunks selected by the switching network. These control functions provide various signaling and maintenance features and also preclude the connection of two or more echo suppressors in tandem.

Because of its high speed, the digital time-shared EST can serve up to 1,680 toll trunks simultaneously. Moreover, it uses less than 10 percent of the power required by analog echo suppressors. Thus, very substantial savings result for the many thousands of trunks that connect to a No. 4 ESS.

Improved 1A Processor for No. 4 ESS

Digital switching and related operations of the No. 4 ESS are controlled by the 1A processor, which is an electronic data processor operating under stored-program control. The basic organization of the 1A processor in the No. 1A ESS (shown in Fig. 2-17) is also applicable to the No. 4 ESS. The main components of the 1A processor and their functions are described in pertinent parts of Chapters 4, 5, and 6, and in the relevent Figs. 4-16, 4-17, 4-18, 5-1, and 6-12.

In initial No. 4 ESS installations, the 1A processor employed ferrite-core memories in program and call stores. The present version utilizes 16K-type MOS integrated circuit memory chips. These provide 262,144 (256K) binary words of memory in each program and call store unit. Thus, a frame of six program and eight call store units contains 1.5 million binary words of program store, and over 2 million words of call store. This compares to a capacity of only 256K words in the same frame space with the original magnetic-core memories.

Moreover, each program and call store now operates at a 700-nanosecond read/write cycle time. This is twice as fast as the original ferrite-core memory and causes a substantial increase in the processing capacity of the 1A processor. In addition, improvements in microprocessor technology enabled the development of a new input/output processor (IOP), which has twice the capabilities of the inital I/O equipment shown in Fig. 2-17. The new *IOP* is compatible with the older I/O equipment, requires only one-half the floor space, and provides more facilities to control additional CRT displays, teletypewriters, and DATASPEED-40 or similar terminals.

The foregoing components of the improved 1A processor and the following associated operation and maintenance centers are illustrated by the block diagrams in Figs. 9-5 and 9-6.

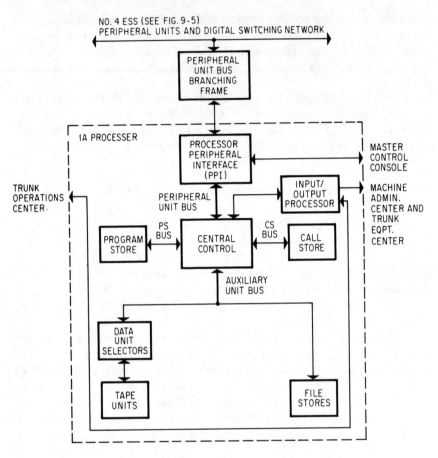

Fig. 9-6. Improved 1A Processor for No. 4 ESS

1. *Trunk operations center (TOC),* which includes trunk test positions, equipment for circuit maintenance and centralized automatic reporting on trunks, and automatic operational testing of the 1A processor.
2. *Machine operations center (MOC)* is the maintenance center for the 1A processor, digital switching network, and the interface terminal equipment. The MOC comprises the *Master Control Console (MCC),* CRT displays, and teletype terminals (see Fig. 9-12).
3. *Terminal equipment center (TEC)* supplements the MOC for transmission maintenance of incoming and outgoing trunks.
4. *Machine administration center (MAC)* handles orders for trunk additions and changes, and directs data changes in the 1A processor with respect to trunk assignments.

No. 4 ESS Digital Switching Network

The digital switching network of the No. 4 ESS, shown in Fig. 9-5, is of the six-stage time-space-time folded type. It consists of two kinds of switching frames: the *time-slot interface (TSI)* and the *time-multiplexed switch (TMS)* frames. The TSI frames provide the initial time-space and the final space-time switching stages. The TMS frames furnish the two-stage space-division switching operations of the network. Four pairs of TMS frames are supplied in a full sized No. 4 ESS office. Each TMS frame has a duplicate or mate frame. Thus, if either unit fails, its mate continues the switching operations and calls are not lost. The complete digital switching network has a total of 64 TSI and 8 (4 duplicated) TMS frames to terminate up to 107,520 trunks and service circuits (1,680 trunks/EST frames × 64 TSI frames).

Each TSI frame interfaces with the echo suppressor terminal (EST) as indicated in Figs. 9-5 and 9-7. The time-shared EST serves up to 1,680 trunks simultaneously, which is the capacity of a TSI frame. The DS-120 PCM signals, which originated in the digital interface frames (DIF) are sent over coaxial cable pairs from the EST to the TSI frame as shown in Fig. 9-7.

The DS-120 format is a 8.192 Megabits per second serial data signal that is generated in the digital interface frames (DIF). Five incoming DS-1 (24-channel PCM or T1 carrier) signals are multiplexed by the DIF into 128 time-slots to construct the DS-120 shape of 128 voice-frequency channels (128 time-slots sampled at 8,000 Hz and encoded at 8 bits per sample = 8.192 Mb/s). The DS-120 designation is used because 120 voice channels are utilized for handling calls and the remaining eight channels are employed for maintenance and control purposes.

The major components of the TSI and TMS frames are illustrated in the block diagrams of Fig. 9-7. The TSI frame, through its peripheal bus interface unit, is controlled over the PUB system by instructions stored in program and call stores of the 1A processor. The timing and control unit of the TSI utilizes the 16.384-MHz timing signal from the network clock. This precise signal furnishes accurate timing for control of time-space and space-time digital switching operations within the TSI frame.

When an incoming toll trunk requests connection to a called area code, the call-processing instructions in program store and the network map memory in call store of the 1A processor are activated. This software initiates a search for an appropriate idle path through the switching network map. Data on specific time-slots and space-switch links of the selected path is sent over the PUB system to the pertinent TSI frame via its peripheral bus interface. This data is interpreted by the timing and control unit. It is then stored in the time-slot memory of the switching and permuting circuits to direct the required digital switching operations. The path selection and net-

Fig. 9-7. Digital Switching Network of TSI and TMS Frames in No. 4 ESS

work map administration functions employed by the 1A processor in the No. 4 ESS are similar in many respects to those in the No. 1A ESS, as described by Fig. 6-7 and its related text in Chapter 6.

Each TMS frame and its mate (which executes the same space-switching operations) has eight switch units in a two-stage switching arrangement as shown in Fig. 9-7. The switch unit in each stage is composed of 16 × 16 array of integrated circuit chips containing crosspoint logic gates. Thus, a TMS frame provides a two-stage 256-input by 256-output space-switching system. The four pairs of duplicated TMS frames, consequently, furnish a total of 1,024 paths in each of the aforementioned 128 time-slots occurring every 125 microseconds. Each switching network connection through the

TMS frames is constituted as a pair of unidirectional paths (one for each transmission direction) in one of the 128 time-slots. Thus, a total of 131,072 (four-wire) network paths (1,024 × 128) can be set up with each path being repeatedly established 8,000 times per second (8,000 Hz).

Referring to Fig. 9-7, PCM signals from the TSI frame are received over coaxial cables by the TMS frame. These signals are then sent through appropriate crosspoint logic gates in the integrated circuit chips of the switch units in the TMS. The TMS only closes the specified space-switch paths for the PCM signals to pass through but does not retime the PCM data. To ensure that network paths are closed at exact intervals, propagation delay must be precisely controlled. This is accomplished by accurate cut-lengths of coaxial cables between TSI and TMS frames, and between the network clock and the respective TSI and TMS frames.

No. 4 ESS Network Clock Unit

The basic timing and synchronization for the No. 4 ESS is furnished by the *network clock* unit. It is the source of precise timing pulses used by the TSI and TMS digital network switching frames and the DIF and LT-1 PCM transmission frames. In addition, the network clock furnishes reference clock pulses for the software in the 1A processor. The major parts of the network clock are four crystal oscillators, four clock chains, and two sets of clock driver (amplifier) circuits as illustrated by block diagrams in Fig. 9-8.

The ultra-precise timing pulses in the network clock are generated by four temperature-controlled 16.384-MHz quartz crystal oscillators. They have a stability of one part in 10 billion (10^{10}) per day. These oscillators are interconnected so that any one can serve as the master oscillator and control the other three, which serve as slaves. In case of an error in the assigned master oscillator's signal, the timing signal from one of the slave oscillators is selected. Likewise, if a slave oscillator becomes faulty, the master oscillator's signal is switched through that slave's clock chain. These error corrections are made without affecting the network clock signals sent throughout the equipment units of the No. 4 ESS.

In Fig. 9-8, the crystal oscillators' sine-wave output signals are formed into square-wave pulses by a circuit in each clock chain. The 16.384 MHz signal from each crystal oscillator feeds all four clock chains. And each clock chain returns a phase-correcting error signal to its associated oscillator. Two clock chains connect to a clock driver unit, which amplifies the timing signals for use within the No. 4 ESS.

With the growth of No. 4 ESS installations, very slight frequency differences or "slip" may occur between two or more No. 4 ESS offices. To

Fig. 9-8. Network Clock and Automatic Synchronization Units in No. 4 ESS (Courtesy of Bell Telephone Laboratories)

preclude this effect, the network clocks in No. 4 ESS offices are monitored and adjusted, if necessary, to the standard 2.048-MHz reference frequency supplied from the Bell System reference frequency nework. This standard frequency originates from an atomic standard located in Hillsboro, MO, near the geographical center of the United States. It is transmitted without regeneration over analog coaxial cable and microwave radio systems to designated timing regions for distribution to No. 4 ESS offices and L5 coaxial cable carrier systems.

The above-mentioned frequency monitoring and adjustments initially were made manually in each No. 4 ESS. The continued growth of No. 4 ESS offices, however, necessitated the development by the Bell Telephone Laboratories of an automatic synchronization unit to automate these frequency checks and adjustments. The automatic synchronization unit, portrayed in Fig. 9-8, performs these functions by means of a digital phase-locked loop circuit. It compares the network clock's signal with the 2.048-MHz Bell System reference frequency and automatically initiates any required frequency corrections to the crystal oscillator or oscillators in the network clock unit.

Trunk Signaling Provisions in No. 4 ESS

Signaling over intertoll and toll-connecting trunks generally involves supervisory and address or control functions. Supervisory signaling, which concerns the initiation and release of call connections and the call status, requires two-way communication. Address or control signaling sends information about the destination of a call, and only needs one-way communication to the called office. The following three types of signaling are provided by No. 4 ESS offices for the aforementioned purposes: dial pulsing (DP), multifrequency (MF) tone pulsing, and common channel interoffice signaling (CCIS).

The dial pulsing (DP) and multifrequency (MF) signaling methods are commonly called "inband" types. They utilize the same voice-frequency (VF) channels used by the calling and called parties. Dial pulse (DP) signaling is transmitted as single-frequency (SF) 2,600-Hz tones. These SF signals are accepted and transmitted by the No. 4 ESS at a nominal 10 pulses per second. MF address signaling comprises combinations of two out of six tones in the 700–1,700 Hz-range. The No. 4 ESS can accept or send one digit by MF every 100 milliseconds. Thus, a 10-digit number can be sent by MF in the same time that is required to transmit the 0 digit by DP. The SF 2,600-Hz tone also is used for supervisory signaling in the above inband signaling methods. Note that with VF or inband signaling, the average time to set up a connection between two electromechanical end offices over two toll-connecting trunks and one toll trunk has been found to be about 10 seconds. With CCIS and ESS offices, including the No. 4 ESS, the call setup time will be only 1–2 seconds. It is anticipated, therefore, that CCIS will replace DP and MF signaling means in the toll network.

Common channel interoffice signaling (CCIS) employs two separate high-speed data links (one for each direction of transmission) for direct communication between No. 4 ESS offices. CCIS also can be used between processors of other stored-program-controlled electronic switching offices. The CCIS method has many advantages in addition to greatly reducing the time needed to establish intertoll connections.

For example, DP and MF signaling is vulnerable to interference from speech and also is susceptible to fraudulent calls. A carrier failure in SF signaling may result in many simultaneous tone-off signals (indicating trunks busy), which can seriously affect switching operations. Furthermore, it is expensive to provide the complex VF signaling equipment at each end of every toll trunk.

DP and MF signaling systems are activated over the E and M leads of their respective trunks. They utilize the same voice paths to be used by the respective calling and called parties. Consequently, the continuity of these transmission paths is assured because they were established as a result of the signaling information transmitted over them. CCIS trunks, however, require continuity checks of their transmission paths, because the voice paths set up for their customers' conversations do not carry signaling information. Separate PCM data links are used between electronic switching offices to convey CCIS data for establishing trunk interconnections. Thus, it is possible for the paths constituted from signaling data to be unacceptable for voice transmission purposes. To preclude this possibility, continuity checks of voice paths are made on every incoming and outgoing trunk to be interconnected in the No. 4 ESS, as explained below.

(For further explanations including schematic diagrams and block drawings of E and M trunk signaling systems incorporating dial pulse (DP), multifrequency (MF), and CCIS methods, the reader is referred to the author's *Basic Telephone Switching Systems,* 2nd Edition, Chapter 4, published by Hayden Book Company, Inc., in 1979.)

CCIS and Call Operations in No. 4 ESS

The major hardware components used in the No. 4 ESS for common channel signaling are represented in Fig. 9-9. They are: CCIS terminals, CCIS continuity check (service) circuits, and CCIS data links. Up to 256 CCIS terminals, 16 in each of 16 terminal grouping frames can be furnished in a No. 4 ESS. Every CCIS terminal is terminated on a specific TSI frame for connection with its voice-frequency data link (VFL). For reliability purposes, both an active and a standby VFL are provided. These CCIS data links currently transmit signaling data at 2,400 bits per second (b/s) over either analog voice channels or digital PCM channels (see Fig. 9-9).

The CCIS data link connects with a CCIS terminal in another electronic switching office or in signal transfer point (STP) centers, because it is not practical to provide separate CCIS data links for each group of trunks between toll centers. Thus, the signaling data for a number of trunk groups is sent to an STP over a single CCIS data link. The STP is a stored-program processor that relays the signaling data to their proper destinations.

Figure 9-9 and the following descriptive summary should help to explain the principal events occurring within the No. 4 ESS for the interconnection of CCIS trunks. In this connection, the CCIS terminal and data link play important roles. The CCIS terminal has three important functions: supervision, addressing, and continuity checking. Supervision includes trunk seizure, called party answer, call disconnect, and abandonment. Addressing refers to the reception of the initial address message (IAM) trans-

DIGITAL SWITCHING NETWORK

CCIS DATA LINK TO OTHER ESS OR STP OFFICES

VFL-1 (ACTIVE)
VFL-2 (STANDBY)

TSI FRAME

T   S
(SEE FIG. 9-7)

TMS FRAME

S   S
(SEE FIG. 9-7)

STAGE   STAGE
#1   #2

4-WIRE OUTGOING CCIS TRUNK

T3, R3
T4, R4

D
C

(2010-Hz TONE) CONTINUITY CHECK TRANSCEIVER

4-WIRE INCOMING CCIS TRUNK

T1, R1
T2, R2

B
A

TO OTHER PERIPHERAL UNITS IN NO. 4 ESS (SEE FIG. 9-5)

NETWORK CLOCK

TERMINAL ACCESS CONTROLLER

16 CCIS TERMINALS

PUB BRANCHING FRAME

1A PROCESSOR

LEGEND:
─○─○─○─ = TIME-DIVISION SWITCHING PATH
─ ─ ─ ─ = SPACE-DIVISION SWITCHING PATH
═══○═══ = 4-WIRE VOICE PATH
VFL-1, VFL-2 = CCIS DATA LINKS
        A = 2-WIRE LOOP FOR CONTINUITY CHECK OF INC. TRUNK
        B = INCOMING CCIS TRUNK PATH IN TSI FRAME
        C = 4-WIRE CONNECTION TO TRANSCEIVER CONTINUITY CHECK CKT.
        D = OUTGOING CCIS TRUNK PATH IN TSI FRAME

Fig. 9-9. CCIS Continuity Checking and Call Switching Paths in No. 4 ESS mitted from the originating office. This address message includes the identification of the incoming toll trunk, and the 7- or 10-digit number of the called party. The continuity check of the selected outgoing trunk is performed by the *continuity check transceiver* as later described.

The CCIS terminal, upon receiving an initial address message (IAM) over the CCIS data link (VFL-1), initiates the necessary operations by the 1A processor and other elements of the No. 4 ESS. Many of these operations, which are described herein, take place almost simultaneously. The CCIS call-handling program first translates the IAM to obtain the TSI frame and time-slot location of CCIS incoming trunk, $T_1, R_1 - T_2, R_2$, shown in Fig. 9-9. Its appearance in the network map memory in call store of the 1A processor is made busy to preclude seizure as an outgoing trunk. Next, the particular TSI frame, on which this incoming trunk appears, is directed to loop together the trunk's two transmission paths, which are designated by A in Fig. 9-9. This permits the distant originating office to make a continuity transmission test of this incoming CCIS trunk over its T_1, R_1 and T_2, R_2 paths.

The digit analysis and routing program in the 1A processor then causes the translation of the address digits (the called number) in order to obtain the outgoing trunk (OGT) routing. Thus, an idle OGT in the trunk routing to the terminating office can be selected in the network map memory of call store. This selected OGT is made busy in the map memory to other outgoing calls.

A continuity check of the selected OGT is next initiated by the CCIS call-handling program. For this purpose, an idle *continuity check transceiver* is connected through the time-division network of the TSI frame to the chosen OGT ($T_3, R_3 - T_4, R_4$) as indicated by path C in Fig. 9-9. Likewise, an IAM is formulated for this selected OGT and sent over the CCIS data link (VFL-1) to the ensuing terminating office.

The equipment in the terminating office proceeds, in a similar manner as previously described for the CCIS incoming trunk, to loop together transmission paths T_3, R_3 to T_4, R_4. At the same time, a 2,010-Hz tone is applied by the continuity check transceiver to path T_4, R_4 and the level of the returned tone on path T_3, R_3 is checked. If the level of the received 2,010-Hz tone is within the specified limit, the continuity check transceiver disconnects the tone. It then transmits a successful continuity test signal over the CCIS data link to the terminating office.

In the meantime, it is assumed that the distant originating office has received a successful continuity check over CCIS incoming trunk $T_1, R_1 - T_2, R_2$. Consequently, the originating office removes its 2,010-Hz tone and sends a continuity message over VFL-1 to the CCIS terminal. Upon the receipt of this message, the CCIS call-handling program will cause the disconnection of loop A from the transmission paths of the CCIS incoming trunk.

Further call-handling operations now await the receipt of both the incoming trunk's continuity message and the successful continuity report on the outgoing CCIS trunk. Upon receiving this information, the 1A processor proceeds to reserve a path in the digital switching network (B to D in Fig. 9-9) between the aforementioned incoming and outgoing CCIS trunks. Next, a continuity completed message for the OGT is sent forward to the terminating office. The call in the No. 4 ESS is presently in the "waiting for address completion" state, since the B to D switching network path has not been connected to the respective incoming and outgoing CCIS trunks.

When the call has been successfully routed through the last terminating office, the latter removes the continuity check loop on its incoming CCIS trunk [see (A) in Fig. 9-9]. The terminating office then transmits an "address complete" signal over the CCIS data link (VFL-1), which is repeated back to the originating office. The CCIS call-handling program, therefore, directs the connection of the incoming and outgoing CCIS trunks to the previously chosen path, B-D, through the switching network. The call, at this point, is in the "waiting for answer" status. Moreover, all transient data pertaining to the call, previously placed in call store of the 1A processor, is erased.

When the called party answers, an answer signal is returned over the CCIS data link for the outgoing CCIS trunk. The CCIS call-handling program, accordingly, will place the call into the talking state. And an answer signal is sent back to the originating office so that charging for the call can be started. If the call cannot be completed because the called line is busy, vacant number, all trunks busy, etc., a clear-forward (CLF) or disconnect message is returned to the originating office instead of the address-complete signal. The receipt of a CLF (disconnect) message enables the originating office to return busy-tone directly to the calling party, instead of from the terminating office encountering the busy-line condition. This procedure makes it possible to instantly break down the various intertoll trunk connections. Thus, they will become immediately available for other calls in lieu of waiting for the calling party to hang up, after an interval of listening to the busy-tone.

Likewise, the CCIS call-handling program will disconnect calls in any state whenever it receives a clear-forward (CLF) message for the incoming CCIS trunk. This CLF (disconnect) signal could result from continuity check failures, trouble conditions, or other reasons. For instance, if the calling party hangs up first at the conclusion of a call, a CLF signal is sent over the CCIS data link to the terminating office. The terminating office responds over the same CCIS data link with a release-guard (RLG) signal, which indicates that the particular trunk is ready to handle the next call. If the called party hangs up first, a clear-back (CB) or hangup signal is transmitted over the CCIS data link to the originating office. A CLF (disconnect) signal is then returned by the originating office, after the elapse of a disconnect timing period in case the calling party is slow in hanging up. The terminating office will acknowledge the receipt of the CLF signal by transmitting back a release-guard (RLG) signal. It indicates that the trunk can handle another call.

Note that the aforementioned digital transmissions over the CCIS data links of the various address and control messages or signals always included the trunk identification data. This identification, as previously mentioned, comprises the TSI frame designation and time-slot location on it of the CCIS incoming or outgoing trunk. Present CCIS data links operate at 2,400 b/s over analog facilities, and about 4000 b/s with digital links. Higher speeds, probably up to 64,000 b/s, can be anticipated over digital facilities in the future.

Mass Announcement and Peripheral Unit Control Frames

A new service feature made possible by the technical architecture of the No. 4 ESS is the *mass announcement system (MAS)*. This optional function can provide various announcement services (local, regional, or nationwide), utilizing the Direct Distance Dialing (DDD) network of the Bell System (see Fig. 1-2) with strategically located No. 4 ESS offices.

Salient types of mass announcement services include public announcements (such as time, weather, news, and sports results), computing *media-stimulated calls*, and cut-through service, which allows a small number of callers access to a "telethon" or call-in sponsor, while most calls are routed to a specified MAS announcement. Public service announcements also may be provided by an audio source other than the MAS frame. These are termed "barge-in" announcements because callers are connected at any point in the announcement cycle after first hearing a cycle of audible ringing.

Media-stimulated calls refer to callers who are aroused through newspaper, television, or radio advertisements to register their opinions on a specific topic by calling certain telephone numbers. For instance, if a "yes" or "no" answer is required, two telephone numbers would be respectively designated. Callers hear an appropriate announcement and the No. 4 ESS counts each call made to the specified number. Several No. 4 ESS offices may be utilized for counting the same media-stimulated calls in a regional area. The resultant totals from all participating No. 4 ESS offices can be instantly transmitted for use by the sponsoring television or radio program.

Requisite equipment for the mass announcement system (MAS), represented in Fig. 9-10, consists of one to eight MAS frames, each with its two disk units, one to four peripheral unit control (PUC) frames (one PUC frame is associated with a maximum of two MAS frames), and two DS-120 links (two coaxial cable pairs). These DS-120 links connect a MAS frame to two dedicated *time-slot interchange switching and permuting circuits (TSI SPC)* in the digital switching network of the No. 4 ESS. The two dedicated TSI frame units (see Fig. 9-7) can provide 1,792 simultaneous call terminations for mass announcement service.

Fig. 9-10. Main Components of Mass Announcement System and Peripheral Unit Control Frames (Courtesy of Bell Telephone Laboratories)

A MAS frame, which comprises two mass announcement units (MAU), has a capacity of 80 30-second audio channel sectors. A maximum of 59 of these time-sectors is used for playback channels to callers. The balance can be employed to store standby audio announcements for subsequent use. Announcements less than 30 seconds still require a 30-second time sector. The greatest possible announcement duration is 5 minutes, which necessitates using 10 30-second sectors. Audible ringing normally is provided so that the caller hears a cycle of ringing tone prior to the announcement. Thus, on the basis of an average 7.5-second waiting time and a

30-second holding time per announcement call, approximately 150,000 calls per hour can be connected to one announcement or to an assortment of announcements on the two TSI frames and the switching and permuting circuits (SPC) dedicated to MAS operations. This caller capacity assumes 90 percent equipment occupancy.

The peripheral unit control (PUC) circuits serve as the control interfaces between the 1A processor in the No. 4 ESS and new service equipment (MAS frames) that will be controlled by the 1A processor. The important components of the PUC frame are depicted in Fig. 9-10. Note that duplicated interfaces with the peripheral unit bus (PUB) of the No. 4 ESS are provided by the bus access circuits in the PUC frame. Likewise, duplex controllers are employed for controlling operations over the extended internal bus (EIB) 0 and 1 to the MAS frames. These means will ensure continued operations in case of failure in a PUC unit.

A unique feature of the PUC frame is that its operations are governed by microprocessors. Therefore, new service features, which may be developed in the future, can be readily added by only changing the program instructions (firmware) of the microprocessors involved. The internal data bus of the PUC circuits is extended outside the PUC frame to serve the MAS frames as well as other equipment that may be installed in the future.

The recording, storing, and subsequent playbacks of mass audio announcements are functions of the moving-head disk storage system and its associated microprocessor control elements. One disk is connected with each of the two mass announcement units (MAU) in a MAS frame, as shown in Fig. 9-10. Announcements are stored concurrently on both disks. Thus, if one disk fails, all announcements are still available. Audio signals for recording and playback of announcements are transmitted over DS-120 links between each MAU and its designated switching and permuting circuits (SPC) in a TSI frame of the No. 4 ESS digital switching network (see Fig. 9-5).

Each DS-120 link (coaxial cable pair) has a capacity of 120 two-way audio and 8 maintenance channels. Normally, 60 channels are used for playback and 14 for recording purposes. Other channels serve as "monitor channels," to verify the integrity of announcements after recording, as well as for maintenance and control requirements. Refer to the previous section, "No. 4 ESS Digital Switching Network," for more particulars on the DS-120 PCM digital format.

The "900" area code would be dialed for the mass announcement service (MAS). Callers usually hear the recorded announcement two or three times and then are automatically disconnected. If desired for certain types of announcements, the No. 4 ESS can be directed to return "answer supervision" to the originating office so that the caller will be charged for the particular call. Announcements can be quickly updated by sponsors or from other sources external to the No. 4 ESS office. A total of 28 announcements per MAS frame may be simultaneously changed.

No. 4 ESS Power Plant

Power requirements for the No. 4 ESS are much greater than for end (Class 5) central offices, such as No. 1 and No. 1A ESS in large cities. Because of its high trunk capacity, total power requirements of a typical No. 4 ESS are usually over 500 kW. In addition, most of this power has to be processed by dc-to-dc converters because of the different closely regulated low voltages needed by the electronic equipment units. Consequently, a 140-volt battery was selected as the primary power source rather than the conventional 48-volt battery. The higher voltage results in lower battery plant cost, as well as considerable reduction in the size and cost of copper cables for distributing dc power to individual equipment frames.

The 24- and 48-volt power needs are supplied by large centralized dc-to-dc converters from the 140-volt battery. Each 24-volt converter furnishes 5-kW output power, and the 48-volt converter has an output rating of 10 kW. The centralized converters contain alarms, circuit breakers, and other features to protect the 24- and 48-volt buses.

The dc voltages required by the No. 4 ESS range from -28 to $+28$ volts. Therefore, modular dc-to-dc converters are installed in each equipment frame to change the 140-volt battery supply to the exact dc voltage levels desired. The major equipment units in the No. 4 ESS are duplicated. Thus, the loss of one power supply converter will not seriously harm system operations. Furthermore, the in-frame converters are rated below 5 amperes output in order to protect the wiring paths in the electronic circuits from damage in case of an accidental short-circuit to a dc voltage bus.

To reduce costs and to simplify maintenance, the dc-to-dc converters have been standardized into 15 types, which are predicated on the following basic power supply circuits:

1. Single-ended switching transistor circuit for output power levels under 30 watts.
2. Two-transistor pulse-width controlled circuit for output power levels from 30 to 350 watts.
3. Silicon-controlled rectifier (SCR) circuit for higher power levels to above 1 kW.

Basic schematic diagrams of dc-to-dc converters utilizing switching power transistors are illustrated in Fig. 9-11. The simplified schematic of a typical single-ended switching transistor converter, designed for power outputs up to 30 watts, is shown in (A) of Fig. 9-11. An important application

Fig. 9-11. Basic Schematic Diagrams of dc-to-dc Converters Utilizing Switching Transistors (Courtesy of Bell Telephone Laboratories)
(A) Single-Ended Switching Transistor Converter
(B) Two-Transistor Pulse-Width Controlled Converter

of this type is to furnish the exact 3-volt power (within 2.90 to 3.10 volts) for the transistor–transistor logic (TTL) chips used in the No. 4 ESS. The following is a brief description of its operation:

Referring to (A) of Fig. 9-11, the control circuit sends a 20-kHz signal through isolation transformer, T1, to switch power transistor, Q1, on and off at that rate. With Q1 switched on, the 140-volt input is applied across the primary of transformer T2. The resultant induced voltage in the secondary of T2 will reverse-bias the diode rectifier, CR1. This causes a series of increasing amounts of current to flow through the primary inductance of

T2. When Q1 is switched off, a voltage is induced in the secondary of T2. It is rectified by CR1 and discharged through the output filter as a decreasing current flow. Regulation of the output voltage is accomplished by the control circuit altering the on-time period of power transistor Q1.

An elementary schematic of a pulse-width controlled converter, designed for 30 to 350 watts output power, is depicted in (B) of Fig. 9-11. Its operation is similar in many respects to the above-described single-ended switching transistor type. For instance, power transistors, Q1 and Q2, are switched on and off alternately by the 20-kHz frequency from the control circuit. The switching operations of Q1 and Q2 produce a pulse-width controlled square-wave voltage across the primary of transformer T1. The resultant induced voltage in the secondary of T1 is rectified by diodes CR1 and CR2 and then filtered to produce the desired dc output voltage. To provide voltage regulation, the control circuit obtains output voltage samples through sampler transformer, T2, in order to control the on-time intervals of power transistors Q1 and Q2.

Dc-to-dc converters cause electromagnetic interference (EMI) emissions that can adversely affect high-speed logic and other operations in the No. 4 ESS. Therefore, the following steps have been taken to minimize EMI radiation in these converters: Bypass capacitors are installed between all power leads to the converters and frame ground, the converter enclosure is shielded, and resistor–capacitor (RC) networks are provided across all fast-switching power transistors. The power supply equipment occupies about 25 percent of the total equipment volume in the No. 4 ESS. It also represents about 15 percent to 20 percent of the total system cost. Thus, the power plant played an important role in the design of the No. 4 ESS.

No. 3 EAX Digital Switching Toll Office

A digital switching toll (Class 4) office also has been developed by GTE Automatic Electric Laboratories, Inc. It is designated No. 3 EAX and has a design capacity of 60,000 terminals for incoming and outgoing trunks. It can switch approximately 350,000 calls during the peak busy hour.

The No. 3 EAX works directly with T1 (PCM) digital trunks. Analog trunks, both two-wire and four-wire, are first converted to the PCM format in a similar manner as described for the No. 4 ESS. The digital switching and associated operations in the No. 3 EAX are directed by its 2B processor, which functions under stored-program control. Higher call-processing speeds are provided by the 2B processor than its predecessor, the 2A processor. Refer to Fig. 8-22 and the related text in Chapter 8 for descriptions of the 2A and 2B processors and their functions.

A basic block diagram of the major components of the No. 3 EAX is represented in Fig. 9-12. The incoming and outgoing digital trunks (T1 lines) terminate on line equipment circuits, one T1 line of 24-voice channels to each such circuit. The line equipment circuits check the synchronization of the T1 lines wich respect to the No. 3 EAX digital network and the availability of time-slots in the time switches. The serial binary digits (bits) output of the PCM speech samples are then combined by the *group equipment* circuits. This combining action results in streams of parallel data for simultaneous transmission to the network system for improved operating efficiency.

The digital switching network of the No. 3 EAX operates in the space-time-space mode, and each space switch is divided into two switching stages. The following is a brief synopsis of the digital switching operations involved, with reference to Fig. 9-12 in the transmission of a PCM speech sample from an incoming to an outgoing digital trunk.

The incoming PCM voice sample received in the line equipment is transmitted to the group equipment circuit for processing. Then it is sent through an assigned path in a space switch to one of the time switches. The

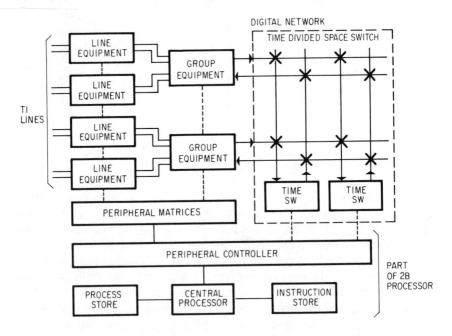

Fig. 9-12. Basic Block Diagram of Major Components of No. 3 EAX Digital Switching Toll Office (Courtesy of GTE Automatic Electric Laboratories, Inc.)

PCM sample is held (delayed) by the time switch until the time-slot period of the selected outgoing digital trunk. At that time period, the PCM sample is transmitted from the particular time-slot in the time switch to the designated outgoing path through a space switch. It proceeds through the space switch to the associated group equipment circuit, that is processing the outgoing digital trunk, to the line equipment circuit in which the outgoing trunk is terminated.

Simultaneously, a PCM speech sample from the outgoing digital trunk is entered in the time switch for subsequent return, at the proper time-slot period, to the originating incoming trunk. The reader can refer to Fig. 9-4 and its related text in this chapter for more detailed explanations applicable to time-division switching operations.

Requisites for Digital Switching End Central Offices

Digital switching systems for local or end (Class 5) central offices incorporate many of the digital transmission and switching techniques that have been described for the No. 4 ESS and No. 3 EAX. They include time-slot interchange and time-multiplexed switches for digital switching functions, and pulse code modulation (T1) carrier systems for digital transmissions. Most toll trunks served by No. 4 ESS and No. 3 EAX offices are derived from either analog FDM carrier systems or digital PCM carrier facilities. These trunks, therefore, have four-wire transmission paths (one two-wire path for each transmission direction). Thus, they can directly interface with a digital switching office.

Subscriber or customer lines connect to central offices over two-wire analog lines, which carry both directions of transmission. Consequently, each customer's line will need a line interface circuit in order to work with the four-wire transmission paths inherent in digital central offices. The transmission elements of a line interface unit, or subscriber line interface circuit (SLIC), are analogous to the hybrid arrangement used for two-wire/four-wire analog trunks that is illustrated in Fig. 9-3.

Several important additional functions must be provided for interfacing a customer's analog line with a digital switching central office. These essential functions are also colloquially known by their Bell System's acronym, "BORSCHT." The seven letters of the acronym, "BORSCHT," and the functions they represent are explained below and portrayed in Fig. 9-13:

B = Battery supply (nominally 48 volts) to power the customer's line equipment (subset) for talking, addressing, and signaling functions.

LEGEND:
  B = BATTERY
  O = OVER-VOLTAGE PROTECTION
  R = RINGING CIRCUIT
  S = SUPERVISION CIRCUITS
  C = CODEC CIRCUITS
  H = HYBRID CIRCUIT
  T = TESTING

SEE FIG. 9-3

Fig. 9-13. BORSCHT Circuit for Subscriber Line Interface with Digital Switching Office (Courtesy of Bell Telephone Laboratories)

O = Overvoltage protection against induced voltages from power lines and lightning strikes. This protective device usually is placed across the T and R conductors of the line prior to its connection to the SLIC from the cable pair.

R = Ringing circuits to control 20-Hz ringing current used to ring the customer's subset. Referring to the simplified schematic diagram in Fig. 9-14, ringing current is applied to the customer's line upon the operation of relay (RG). The 20-Hz ringing current flows through the winding of the (RR) relay over the R line conductor to the customer's subset and back over the T conductor to ground in the ringing supply circuit. When the called party answers (off-hook), relay (RR) operates, thereby releasing relay (RG). These operations disconnect the ringing circuit and close through the T and R leads for transmission.

S = Supervision of the customer's line to detect service requests (off-hook) and disconnection (on-hook) states. Likewise, the detection of direct-current (dc) flow in the subscriber's loop circuit (off-hook) and dial pulses (periodic interruptions of the line's continuity).

C = Codec equipment, which comprises A/D (encoder) and D/A (decoder) converters with their associated low-pass filters. See Fig. 9-3 for further details.

H = Hybrid transformer and balance network circuit to interface the 2-wire line circuit (used for both directions of transmission) with the four-wire digital switching network, which provides separate paths for each transmission direction (see Fig. 9-3). An active hybrid circuit, employing integrated circuits and a voltage divider system, may be used instead of the hybrid transformer to minimize space requirements. But this could necessitate additional reliability provisions.

T = Testing access to the customer's line. This task, as in the case of the "R" (ringing circuit) function, is usually accomplished by interrupting the T and R conductors of the line through relay contacts. Refer to the (TEST) relay in Fig. 9-13.

Digital Telephone Set Contingencies

A digital telephone set incorporating the aforementioned "BORSCHT" functions would cost a great deal more than analog telephones. Furthermore, a four-wire line would be needed to the digital switching office in addition to a power supply at the customer's premises. Moreover, each extension station likewise would require a digital telephone set.

It is possible, however, to utilize two wires for connecting digital telephones to the digital switching office or to a remote digital terminal. But this method will result in a "ping-pong" type of transmission on the two-wire line. This technique also could cause degradation in the transmission quality of the associated PCM (T1) carrier system.

The above and related cogent reasons indicate the need for continuing the use of present analog telephones pending further technological developments. Consequently, subscriber line interface circuits (SLIC) have been included as integral parts of Class 5 digital switching offices. The SLIC, under microprocessor control, can adequately furnish the above-mentioned "BORSCHT" facilities for customers served by digital end central officers.

It is more economical, however, to provide the analog/digital conversion and other "BORSCHT" features at a nearby remote terminal or line-concentrator unit located in the customer's community. This arrangement, which can serve numerous customers, is similar in many ways to the pair gain method described below. The remote terminal is controlled and monitored by microprocessors. Information and status data (on-hook, off-hook, conversation, etc.) detected by the microprocessors are sent with associated identifying data to the central processing unit (CPU) in the digital switching office. Whenever a status check or other processing action is required, the CPU immediately notifies the pertinent microprocessor in the remote terminal.

Pair Gain System for Digital End Offices

Present analog telephones, as previously discussed, will require line interface equipment to function with digital switching offices. In this connection, an improved *pair gain* method, utilizing the T1 carrier (PCM) system, has been developed by the Bell System and other telepone operating organizations. It furnishes better and more economical facilities for serving customers of digital switching offices over the existing subscriber cable plant, especially in urban and suburban areas. This pair gain method also mitigates requirements for new subscriber cables by increasing the number of customer lines that can be served by the existing cable plant.

The main components of the pair gain system are: remote terminals, which include a channel unit for each customer's cable pair, and T1 lines (24-channel PCM carrier circuits) to connect the remote terminals directly to the digital switching equipment in a Class 5 digital office. Battery power and ringing current for remote terminals are provided in a similar manner as for community dial offices (CDO) or large PBX installations of equivalent capacity.

Figure 9-14 is a functional block diagram of the pair gain arrangement evolved by Bell Telephone Laboratories, Inc. It employs a subscriber loop carrier, designated SLC-96, which can serve from 24 to 96 customer lines depending on the concentration. For example, a 2:1 concentration enables 48 customers to share the 24-channel capacity of one T1 line. If all 96 customers are to be served without concentration, five T1 lines are furnished. They include one T1 line for protection, which is automatically switched into service in case of a failure in any of the other T1 lines.

The simplified block diagram in Fig. 9-14 shows the connection of the SLC-96 *pair gain* system to 48 customer lines served by a remote terminal. Each customer's line interfaces with a channel unit that functions in a similar fashion as shown in Fig. 9-12 for the BORSCHT circuit. The analog signals from customers' lines are converted to PCM samples by the channel units and then multiplexed into 24-channel digital signals by the transmit/receive unit (TRU). The TRU also assigns the digital signal, from a customer's line, a specific time-slot in order to differentiate it from the other 23 PCM signals that are being multiplexed. The line interface unit (LIU) shapes the digital signals for transmission over the T1 line, and also terminates the T1 line for both transmission and receiving directions.

Fig. 9-14. Major Components of SLC-96 Pair Gain System (Courtesy of Bell Telephone Laboratories)

LEGEND:
← = 2-WIRE TRANSMISSION PATH
TRU = TRANSMIT/RECEIVE UNIT
TAU = TIME ASSIGNMENT UNIT
T1 LIU = T1 LINE INTERFACE UNIT

The time assignment unit (TAU) is used to increase the number of customers served by one T1 line from 24 to 48. For this purpose, the TAU collects the digitized signals or pulse groups from the TRU before they are transmitted through the LIU. The TAU allots these pulse groups, which represent customer calls, to available digital channels in the 24-channel T1 line. This technique also may be defined as selective matching of signals for transmission with available timeslots. Communication between time assignment units in *remote terminals* and the digital switching office are provided by a two-way data link. This data link also handles alarms, customer line test data, and T1 line protection switching.

Digital Switching Systems for Class 5 Offices

Advancements in semiconductor technology, particularly the development of low-cost microprocessors and large-scale integrated (LSI) circuits, have contributed to improved economical designs of digital switching systems for Class 5 offices. For instance, one LSI chip can perform many complex functions, such as A/D and D/A conversions, and adjust the balance network of active hybrid circuits in line interface units. Likewise, many microprocessors, each with its related memory, can be utilized to control various call processing, path selection, and other functions in the digital switching system. This method greatly reduces the work of the large central processor unit (CPU) usually employed for these purposes. Consequently, call-handling capacities of digital switching systems could be substantially increased at low cost.

Digital switching systems for end offices are usually designed to work with T1 lines, or synonymous PCM carrier circuits, as previously described for the No. 4 ESS. Their terminals interface with customers' cable pairs either at terminations in the central office or at remote terminal sites. The time-space-time type of digital switching network generally is utilized with variations on the stored-program control process.

A number of leading American and foreign laboratories including equipment manufacturers are presently engaged in developing improved digital switching systems suitable for both small and large Class 5 offices. The following are succinct synopses of some pertinent digital switching systems now under development or undergoing field evaluations and service tests.

GTD-5 EAX

The GTD-5 EAX, developed by GTE Automatic Electric Laboratories, Inc., operates under stored program control to serve Class 5 offices. It also can function with remote terminals via T1 lines in an analogous manner to the SLC-96 pair gain method described in Fig. 9-14. The small base type of GTD-5 EAX has a capacity of 500 to 20,000 lines and 65,000 busy-hour call attempts to serve rural and small suburban communities. The large base model is designed for metropolitan cities and expanding suburban areas. It can handle 2,000 to 150,000 lines and has a capacity of 360,000 busy-hour call attempts.

A significant feature in the GTD-5 EAX is the use of the distributing processing system. There are over 300 microprocessors, instead of only a large central processor unit (CPU), for controlling call processing and associated digital switching operations. Figure 9-15 is a simplified block diagram of the major hardware items of the GTD-5 EAX base unit, including the peripheral equipment, digital switching network, and the central control equipment.

Customer lines, analog trunks, and service circuits connect to the GTD-5 EAX through analog facility interface (AFI) units. T1 lines from remote terminals and digital trunks connect through the digital facility interface (DFI) circuits of the peripheral equipment. The AFI units provide, on a per-line or per-trunk basis, analog to PCM conversions to interface with the digital switching network, and one AFI circuit can furnish a maximum of 192 PCM channels, in each transmission direction, to the switching network. Likewise, a DFI or digital trunk interface element provides terminations for 8 DS-1 circuits (8 × 24 PCM channels) or a total of 192 PCM channels. These interconnections between the peripheral equipment and the digital switching network are represented as PCM channels in Fig. 9-15.

The digital switching network of the GTD-5 EAX is of the time-division multiplexed type and consists of three (time-space-time) stages. This three-stage network is divided into time switch and space switch modules. Every time-switching module handles the voice-channel outputs from four

Fig. 9-15. Simplified Block Diagram of Major Equipments of GTD-5 EAX System (Courtesy of GTE Automatic Electric Laboratories, Inc.)

facility interface (AFI or DFI) units for a total of 768 (192 × 4) PCM channels. The space switch is a fully duplicated module that interconnects the outputs of 32 originating time switches to the inputs of 32 terminating time switches. A maximum of 64 time switches and 4 space switches normally may be equipped.

The central control processing equipment, illustrated in Fig. 9-15, comprises an administrative processor complex (APC) and up to seven telephony processor complexes (TPC), depending on the number of lines and traffic requirements. All of these equipment components are duplicated. The APC handles administrative and maintenance functions, whereas the TPCs perform call-processing operations. In addition, a peripheral processor (PP) is associated with each group of four AFI and four DFI units, and their connected time switch in the switching network. These peripheral processors (PP) and the aforementioned APC and TPC processors in central control contain duplicated dedicated memories. These memories include protected program instructions and data, as well as unprotected or temporary data pertaining to calls in progress and related network connections.

Data and pertinent instructions from the peripheral processors (PP) are sent via the message distributor circuit to the APC and relevant TPC processors in central control. These central control processors then control and assign the tasks to be performed within the entire base unit of the GTD-5 EAX. The *ADC* also interfaces to the administrative control and display complex, and likewise with an input/output module for magnetic tape data. The administrative control and display unit provides maintenance personnel with system-operating information and emergency configuration control whenever required.

NEAX 61

The NEAX 61 is a digital telephone switching system, recently developed by Nippon Electric Co., Ltd., which also is suitable for Class 4, Class 5, and tandem office installations in the United States. Its design capacity is 100,000 lines and 13,000 trunks for Class 5 offices, and 60,000 toll switching trunks for Class 4 installations. The call-handling capacity is approximately 700,000 busy-hour call attempts. The NEAX 61 utilizes many microprocessors to provide distributed processing with stored-program control. It employs PCM type of time-division switches. The NEAX 61 structure consists of four major parts or subsystems that are represented by the block diagrams in Fig. 9-16, namely, application, switching, processor, and maintenance and administration.

The application subsystem consists of several types of service interface modules for the control of various terminal circuits in their operations with other subsystems of the NEAX 61. These types include interface modules for customer lines, analog and digital trunks, and PCM carrier channels (T1 lines) to remote terminals serving customers. The customer's line interface module incorporates an equivalent BORSCHT circuit that is portrayed in Fig. 9-13. All interface modules contain terminal and interface circuits, a duplicated controller, and a duplicated primary multiplexer/demultiplexer (PMUX). The PMUX provides 128 PCM channels (120 voice and 8 for control purposes) that connect with the secondary multiplexer/demultiplexer (SMUX) in the switching system as shown in Fig. 9-16.

The switching subsystem is a four-stage folded (time-space-space-time) type of time-division multiplexed switching network. It contains up to 22 duplicated network modules or frames, each controlled by a call proces-

APPLICATION SUBSYSTEM   SWITCHING SUBSYSTEM

SERVICE INTERFACE MODULES

NETWORK MODULE #1

CUSTOMERS
2W LINES

ANALOG
2W
TRUNKS

TI (PCM)
CARRIER
SYSTEM

ANALOG
LINES — PM / UX

REMOTE
TERMINALS — PM / UX

ANALOG
TRUNKS — PM / UX

DIGITAL
TRUNKS — PM / UX

SMUX

TIME
SWITCH
#1

SPACE
SWITCH

TO AND
FROM
OTHER
SWITCHING
MODULES

TO AND FROM
OTHER SERVICE
INTERFACE
MODULES

SMUX

TIME
SWITCH
#6

SERVICE
CKT.

SPEECH PATH
CONTROLLER

PROCESSOR SUBSYSTEM

CALL
PROCESSOR — MEMORY

INTEGRATED BUS

TO OTHER CALL
PROCESSORS

TO
OTHER CALL
PROCESSORS

MAINTENANCE & ADMINISTRATION SUBSYSTEM

OPN. & MTCE.
PROCESSOR — MEMORY

INPUT/OUTPUT BUS

TO
OTHER I/O
CONTROLLERS

LEGEND:
= 4W TRANSMISSION PATH
PMUX = PRIMARY MULTIPLEXER/DEMULTIPLEXER
SMUX = SECONDARY MULTIPLEXER/DEMULTIPLEXER
OPN.& MTCE.= OPERATION AND MAINTENANCE
                 PROCESSOR
I/O CLR.   = INPUT/OUTPUT CONTROLLER

I/O CLR — CRT

I/O CLR — TTY

I/O CLR — MAG. TAPE

Fig. 9-16. Abridged Block Diagram of NEAX 61 Digital Telephone Switching System (Courtesy of Nippon Electric Co., Ltd.)

sor. A speech path controller in each network module is the control memory of the four switching stages. It establishes paths through the network modules under control of its associated call processor. Service circuits, which direct the operations of the space switches, are also controlled by the speech path controller.

The processor subsystem involves numerous call processors (each with individual memories), the operator and maintenance processor, and the interconnecting integrated bus between the various elements of the processor subsystem. Control data is passed between call processors over the integrated bus to direct speech path connections between network modules. Maintenance routines and related operations are executed by the operation and maintenance processor (OMP). The OMP also governs the operation of call processors for test purposes and to provide manual control access of individual modules.

The maintenance and administration subsystem comprises different types of input/output controllers that connect to the *OMP* over an input/output (I/O) bus. I/O controllers also are included for magnetic tape units used for AMA operations, as well as for the CRT terminals and teletypewriters employed for communicating with maintenance personnel.

ITT System 12

A new generation of digital switching systems, designated ITT System 12, has been developed by laboratories of the International Telephone and Telegraph Corporation in the United States and Europe. These new systems include the ITT 1210 (formerly known as the DSS-1 of ITT North Electric Co.), ITT 1230 types, which are suitable for rural and small-to medium-size Class 5 offices, the ITT 1220 for Class 4 toll and tandem offices, and type ITT 1240 for large Class 5 digital offices in metropolitan localities.

The ITT System 12 digital switching systems employ stored program control (SPC) and the distributed processing method. This procedure utilizes many microprocessors with small memories rather than a central processor unit (central control and program and call stores used in the 1A processor) to control the various call-processing and digital switching operations. The use of microprocessors and LSI logic circuits has significantly increased line and trunk capacities with substantial reductions in hardware requirements in these systems.

The design capacity of the ITT 1220 digital toll office is 50,000 trunk terminals for Class 4 operations and up to 600,000 busy-hour call attempts. Its digital switching network is the time-space-space-time type, which is duplicated for reliability. The ITT 1240 model, planned for large size end (Class 5) offices, can handle up to 100,000 customer lines or 60,000 trunks. Microprocessors, with only the required amount of memories, provide direct hardware control of interface circuits with customer lines. For example, one microprocessor is associated with every 60 customer lines in a line interface circuit. These line interface circuits interconnect with the digital switching network through a terminal interface unit that provides digital signals on a PCM basic. Trunks are connected to the switching network in a similar fashion, and a microprocessor is related with each group of 30 trunks.

No. 5 ESS

An improved digital system for end central offices, termed No. 5 ESS, is currently under development by the Bell Telephone Laboratories, Inc. It is understood that operations of the No. 5 ESS will be executed by very high speed network processors in accordance with stored programs. A separate common control computer would be employed, as needed, to update the data in the network processors. The various time and space switching functions would be accomplished simultaneously by designated network processors.

If it is assumed that a network processor can connect to 8,000 customer lines, an average of 4,000 lines may be served by each processor. On this basis, approximately 250,000 customers could be served by 64 network processors in a No. 5 ESS. Furthermore, the No. 5 ESS will be able to provide many new service features over the usual wire pair to a customer's premises. These services could include, for example, simultaneous voice and low digital data rate (300 bits/second) in both directions for computer terminal usage and security and alarm functions.

Questions

1. What is the theoretical sampling rate for changing analog voice signals to the digital form? What sampling rate has been standardized for voice-frequency signals?
2. Name the major advantages of digital signals over analog. What is one practical disadvantage of digital transmission?
3. What functions are performed by the hybrid circuit and its balance network?
4. Explain the equipment items of the No. 4 ESS that interface with incoming and outgoing toll trunks. What are the other major hardware components of the No. 4 ESS?
5. Describe the digital switching network of the No. 4 ESS. What is the design capacity?
6. Name the functions performed by the network clock unit in the No. 4 ESS. What is the frequency and stability of its crystal oscillators? What is the standard reference frequency?
7. What is the primary power source in the No. 4 ESS and why was it selected? How are the required dc voltages obtained at equipment frames?
8. Name the chief components of the GTE No. 3 EAX digital switching toll office and indicate its design capacity.

9. What acronym designates the functions performed by a subscriber's line interface unit in a digital switching Class 5 or end central office? Explain these functions.
10. What improved technological feature has been designed in the new generation of digital switching systems under development for Class 5 offices? List two or more of these new systems and their estimated line capacities.

Index